Applications of Lie's Theory of Ordinary and Partial Differential Equations

Applications of Lie's Theory of Ordinary and Partial Differential Equations

Lawrence Dresner

Oak Ridge, Tennessee, USA

CRC Press
Taylor & Francis Group
Boca Raton   London   New York

CRC Press is an imprint of the
Taylor & Francis Group, an **informa** business

CRC Press
Taylor & Francis Group
6000 Broken Sound Parkway NW, Suite 300
Boca Raton, FL 33487-2742

© 1998 by Taylor & Francis Group, LLC
CRC Press is an imprint of Taylor & Francis Group, an Informa business

No claim to original U.S. Government works

ISBN-13: 9780750305310

Visit the Taylor & Francis Web site at
http://www.taylorandfrancis.com

and the CRC Press Web site at
http://www.crcpress.com

Contents

Preface

Lie's group theory of differential equations, which is more than a hundred years old, has had a rebirth in the last twenty years as people have again begun to appreciate its virtues, namely: (1) that it unifies the many *ad hoc* methods known for solving differential equations, and (2) that it provides powerful new ways to find solutions. The theory has applications to both ordinary and partial differential equations and is not restricted to linear equations. What is needed now is a short, simple, introductory text to spread the word.

I am not satisfied that recent books fill this need. Like many modern mathematics books, they are written in an encyclopedic style that makes them hard to read. In striving for rigor and generality right from the outset, their authors make the underlying ideas not only difficult to understand but sometimes even difficult to identify. The book that we want should be, to borrow a phrase of Jacques Barzun's, 'simple and direct.'

A short monograph similar to the well-known Methuen monographs seems perfect for this purpose. These monographs have roughly 150–200 9 cm × 14 cm pages of 10-point type and are intended as an introduction to a single subject. Their purpose, for which I esteem them, is to get the reader 'up and running' as quickly as possible. To do this here, I have sacrified not only pantological completeness but rigor and generality in favor of clarity and immediacy of understanding.

To work with less than complete rigor and generality, even if only for a time, is to swim against the current of modern mathematics. But there is good authority for it in the writings of a mathematician of unassailable reputation, George Polya. In his book _How to Solve It_, Polya calls attention to the value of what he terms incomplete proofs, which he describes 'as a sort of mnemotechnic device ... when the aim is tolerable coherence of presentation and not strictly logical consistency.' After all, he says, 'the facts must be presented in some connection and in some sort of system, since isolated items are laboriously acquired and easily forgotten. Any sort of connection that unites the facts simply, naturally, and durably, is welcome here ... proofs may be useful, _especially simple proofs_' (italics added). One goal of this book, then, is to explain matters so that the the reader can see them, as Polya puts it, 'at a glance'.

I do not think that Polya intended for his phrase 'at a glance' to imply 'without much thought'. Instead I think he meant that with study and thought, a clear, unified picture would emerge with which the reader could thereafter think about the subject fluently. In this connection, it is probably worth taking as a _caveat_ to Polya's dictum Einstein's quip that 'things should be made as simple as possible, but not simpler'. Consequently, some sections of this book are harder than others. As a help to the reader, I have marked the more difficult sections with a star. These sections may be omitted on a first reading; the reader may thereby gain an overview of the subject before coming to grips with the more complicated details.

A second goal of this book is to give the reader some hint of what to do once he has found a group under which his differential equation is invariant. As a practical matter, what to do next is as important in finding a solution as finding a group in the first place. It has been my observation that in practice the most frequently occurring groups are stretching (affine) groups, translation groups, or combinations of the two. Often the simpler equations are invariant to an entire family of such groups. Such great symmetry (as group invariance is often termed) allows some rather general theorems about the solutions to be derived, and such theorems are discussed and their use illustrated.

There are problems at the end of each chapter which I have included as an aid to learning. It has been said of piano playing that 'nobody ever learned to play the piano by reading books on how to play the piano'. The same is true of mathematics: reading a book is not enough; doing is essential to mastery. Therefore, I exhort the reader to solve the problems. I have also included my own solutions to the problems. If the reader gets stuck, I recommend that he read just far enough in my solution to get past his block and then continue his own efforts. In this way, he can maximize the instructive value of the problems.

I have written this book in comparative collegial isolation, following my own thoughts wherever they led me. Nevertheless, I owe a debt of thanks to Professor Barbara Schrauner for encouraging me over the years to remain engaged with the subject of Lie groups. I should also like to express my thanks to my editor, Mr Jim Revill of Institute of Physics Publishing, for bringing this book into the light of day, and to his anonymous referees, whose enormous critical labor helped greatly to improve the book.

Lawrence Dresner

Oak Ridge, Tennessee
July 1998

Conventions Used in this Book

As already mentioned in the Preface, as a help to the reader, I have marked the more difficult sections with a star. These sections may be omitted on a first reading; the reader may thereby gain an overview of the subject before coming to grips with the more complicated details.

Interspersed throughout the text are examples, which are introduced by the word 'Example' and end with the symbol ■.

There are notes at the end of most chapters that contain additional discussion of points raised in the main text.

Frequent use is made of the subscript notation for partial derivatives. Thus the ordinary diffusion equation $\partial c / \partial t = \partial^2 c / \partial z^2$ is written $c_t = c_{zz}$ for the sake of economy in typesetting. Similarly Newton's notation for the ordinary derivative is often used in place of Leibniz's; thus for the function $y(x)$, $\dot{y} = dy/dx$ and $\ddot{y} = d^2y/dx^2$.

Dedication

To the memory of my father
Max Dresner
1905–1973
and
my father-in-law
Ben Hershman
1908–1995

*Now go, write it before them in a table, and note it in a book, that it may be
for the time to come for ever and ever.*

Isaiah 30: 7–8

1

One-Parameter Groups

1.1 Groups of Transformations†

We begin by considering sets of transformations

$$x' = X(x, y; \lambda) \qquad (1.1.1a)$$
$$y' = Y(x, y; \lambda) \qquad (1.1.1b)$$

that depend on a continuous parameter λ. Any particular value of λ determines one transformation of the set. Each transformation of the set may be looked upon as mapping any point (x, y) in the unprimed plane into an image point (x', y') in the primed plane. An example of such a set of transformations is

$$x' = x \cos \lambda - y \sin \lambda \qquad (1.1.2a)$$
$$y' = x \sin \lambda + y \cos \lambda \qquad (1.1.2b)$$

These transformations are called rotations because the position of the image point (x', y') can be determined by rotating the radius vector to the source point (x, y) counterclockwise through an angle λ. (This can easily be seen by setting $z = x + iy$ and noting that Eqs. (1.1.2a, b) are the real and imaginary parts of the equation $z' = ze^{i\lambda}$.)

† The transformations (1.1.1) are *point transformations*, which are the only transformations dealt with in this book. Other, more general transformations have been studied by Lie and by others, but their consideration is beyond the scope of this introductory book. Some references to them are given in the epilog.

The rotations (1.1.2a, b) have certain properties not possessed by every set of transformations. In the first place, two rotations carried out in succession are equivalent to another single rotation. Thus, if we apply (1.1.2a, b) to the source point (x', y') using the value λ' of the angle, we obtain an image point (x'', y'') that can be obtained from the original source point (x, y) by a single rotation through the angle $\lambda'' = \lambda + \lambda'$.

If $\lambda' = -\lambda$, then the second image point (x'', y'') coincides with the source point and the first two rotations are inverses of one another. Furthermore, their resultant is the identity transformation for which the source and image points coincide.

When a set of transformations has these three properties, namely, (1) two transformations carried out in succession are equivalent to another transformation of the set;[1] (2) there is an *identity* transformation for which the source and image points coincide; and (3) each transformation has an *inverse*; the set is said to form a *group*. The property (1) is often called the *group property*. In this book, we shall only consider groups of transformations.

1.2 Infinitesimal Transformations

Let us denote by λ_0 the value of λ for which the transformation (1.1.1a, b) becomes the identity transformation, i.e. the one for which $x = X(x, y; \lambda_0)$ and $y = Y(x, y; \lambda_0)$. If we expand (1.1.1a, b) in a Taylor's series around the point $\lambda = \lambda_0$, we find to first order,

$$x' = x + \left(\frac{\partial X}{\partial \lambda}\right)_{\lambda=\lambda_0} (\lambda - \lambda_0) + \dots \qquad (1.2.1a)$$

$$y' = y + \left(\frac{\partial Y}{\partial \lambda}\right)_{\lambda=\lambda_0} (\lambda - \lambda_0) + \dots \qquad (1.2.1b)$$

The partial derivatives $(\partial X/\partial \lambda)_{\lambda=\lambda_0}$ and $(\partial Y/\partial \lambda)_{\lambda=\lambda_0}$ are functions of x and y, which we denote, respectively, by $\xi(x, y)$ and $\eta(x, y)$. For sufficiently small $\lambda - \lambda_0$ we can then write the coordinates of the image point (x', y') as

$$x' = x + \xi(x, y)(\lambda - \lambda_0) \qquad (1.2.2a)$$

$$y' = y + \eta(x, y)(\lambda - \lambda_0) \qquad (1.2.2b)$$

where we have now dropped terms of second and higher order in $\lambda - \lambda_0$. This first-order transformation is known as the *infinitesimal transformation*.

If we apply the infinitesimal transformation $(1.2.2a, b)$ to the point (x', y') we obtain the coordinates of the second image (x'', y'')

$$x'' = x' + \xi(x', y')(\lambda - \lambda_0) \qquad (1.2.3a)$$

$$y'' = y' + \eta(x', y')(\lambda - \lambda_0) \qquad (1.2.3b)$$

Because of the group property, the point (x'', y'') is also an image of the original source point (x, y). Continuing in this way, we can advance by infinitesimal steps from the source point (x, y) to its remote images.

The infinitesimal transformation $(1.2.2a, b)$ is Euler's finite-difference algorithm for solving the coupled differential equations

$$\frac{dx}{\xi(x, y)} = \frac{dy}{\eta(x, y)} = d\lambda \qquad (1.2.4)$$

The trajectory through the source point (x, y) defined by Eq. (1.2.4) is the locus of all the images of the source point (x, y). This locus is called the *orbit* of (x, y); it is, clearly enough, the orbit of any other point on the locus, so we may speak unambiguously of the orbits of the group.

We shall call the functions $\xi(x, y)$ and $\eta(x, y)$ the *coefficients* of the infinitesimal transformation $(1.2.2a, b)$. Their usefulness arises from the fact that often we have an explicit formula for them whereas we do not often have an explicit form for the functions X and Y. The latter, of course, can be obtained from $\xi(x, y)$ and $\eta(x, y)$ if we can integrate the differential equations (1.2.4).

Example: For the group $(1.1.2a, b)$, $\lambda_0 = 0$ and $\xi(x, y) = -y, \eta(x, y) = x$. Thus (1.2.4) becomes

$$-\frac{dx}{y} = \frac{dy}{x} = d\lambda \qquad (1.2.5)$$

The first of this pair of equations can be integrated to give $x^2 + y^2 = a^2$, where a^2 is a constant of integration. Then the second equation of the pair (1.2.5) can be written

$$(a^2 - y^2)^{-1/2}\,dy = d\lambda \qquad (1.2.6)$$

Eq. (1.2.6) can be integrated to give

$$\arcsin\left(\frac{y}{a}\right) = \lambda + b \qquad (1.2.7)$$

where b is a second constant of integration. It follows then that

$$x = a\cos(\lambda + b) \qquad (1.2.8a)$$
$$y = a\sin(\lambda + b) \qquad (1.2.8b)$$

which is equivalent to (1.1.2*a, b*) if we take

$$x = a\cos b \qquad (1.2.9a)$$
$$y = a\sin b \qquad (1.2.9b)$$

as the coordinates of the source point.■

1.3 Group Invariants

A *group invariant* is a function $u(x, y)$ whose value at an image point is the same as its value at the source point:

$$u(x', y') = u(x, y) \qquad (1.3.1a)$$

or

$$u(X(x, y; \lambda), Y(x, y; \lambda)) = u(x, y) \qquad (1.3.1b)$$

Thus it is constant along an orbit although it may have different values on different orbits.

The left-hand side of Eq. (1.3.1*b*) is a function of λ while the right-hand side is not. If we differentiate Eq. (1.3.1b) partially with respect to λ and then set $\lambda = \lambda_0$, we find

$$\xi(x, y) u_x + \eta(x, y) u_y = 0 \qquad (1.3.2)$$

where we have used the conventional subscript notation for the partial derivatives (i.e. we have abbreviated $\partial u/\partial x$ as u_x, etc.).

The characteristic equation† associated with the linear, first-order partial differential equation (1.3.2) is

$$\frac{dx}{\xi(x, y)} = \frac{dy}{\eta(x, y)} \qquad (1.3.3)$$

and the general solution of Eq. (1.3.2) is an arbitrary function of an integral of Eq. (1.3.3). This arbitrary function is then the most general group invariant.

This result is not surprising because the trajectories defined by Eq. (1.3.3) are the orbits of the group (cf. Eq. (1.2.4)). An integral of Eq. (1.3.3) is a function that is constant along any of the trajectories; thus it is a function that is constant on any orbit of the group. For the rotation group (1.1.2*a*, *b*) such an integral is $x^2 + y^2$ so that any function of it, $F(x^2 + y^2)$, represents that group's most general group invariant.

1.4 Invariant Curves and Families of Curves

An *invariant curve* C is one whose points, considered as source points, map into other points of the curve C for all transformations of the group. Thus C must either be an orbit or a locus on which the infinitesimal coefficients $\xi(x, y)$ and $\eta(x, y)$ simultaneously vanish.

A one-parameter family of curves can be represented parametrically by the equation

$$\phi(x, y) = c \qquad (1.4.1)$$

---

† The theory of linear, first-order partial differential equations is summarized in Appendix A. The reader unfamiliar with it must now consult Appendix A because the remainder of this chapter depends heavily on this theory.

where ϕ is the function defining the family and c is a parameter that labels different curves of the family. The family is said to be invariant if the image of each curve of it is another curve of the family. The condition for this is that for any fixed value of λ the image points (x', y') satisfy

$$\phi(x', y') = \phi(X(x, y; \lambda), Y(x, y; \lambda)) = c' \qquad (1.4.2)$$

when the source points (x, y) satisfy Eq. (1.4.1). Here c' is a parameter different from c whose value depends on c and λ. If we now differentiate Eq. (1.4.2) partially with respect to λ and then set $\lambda = \lambda_0$, we find

$$\xi(x, y)\,\phi_x + \eta(x, y)\,\phi_y = \left(\frac{\partial c'}{\partial \lambda}\right)_{\lambda=\lambda_0} \qquad (1.4.3)$$

The right-hand side of Eq. (1.4.3) is a function only of c; call it $F(c)$. In view of Eq. (1.4.1), Eq. (1.4.3) can be written

$$\xi(x, y)\,\phi_x + \eta(x, y)\,\phi_y = F(\phi) \qquad (1.4.4)$$

The representation (1.4.1) of the family of curves is not unique and any other representation

$$\psi(x, y) = c_1 \qquad (1.4.5)$$

for which ψ is a function of ϕ, i.e. for which

$$\psi = G(\phi) \qquad (1.4.6)$$

is equivalent to the representation (1.4.1). It follows from Eq. (1.4.4) that

$$\xi\psi_x + \eta\psi_y = (\xi\phi_x + \eta\phi_y)\frac{dG}{d\phi} = \frac{dG}{d\phi}F(\phi) \qquad (1.4.7)$$

Since we can choose the function G at our pleasure, we could choose, for example,

$$G(\phi) = \int \frac{d\phi}{F(\phi)} \qquad (1.4.8)$$

for which choice the right-hand side of Eq. (1.4.7) becomes 1:

$$\xi\psi_x + \eta\psi_y = 1 \tag{1.4.9}$$

The choice of the function F is up to us; different choices correspond to different ways of parametrizing the family of invariant curves; the family in its entirety remains the same.

Example: Let us continue the example of the group (1.1.2a, b) for which $\xi = -y$ and $\eta = x$. The characteristic equations of the linear partial differential equation (1.4.9) are

$$-\frac{\mathrm{d}x}{y} = \frac{\mathrm{d}y}{x} = \mathrm{d}\psi \tag{1.4.10}$$

These equations have two independent integrals, and the most general solution for ψ is obtained by equating to zero an arbitrary function of these integrals. Integrating the first of this pair of equations, we find as before, $x^2 + y^2 = a^2$, where a^2 is a constant of integration. If we now substitute $(a^2 - y^2)^{1/2}$ for x and integrate the second equation we find

$$\psi - \arcsin\left(\frac{y}{a}\right) = b \tag{1.4.11}$$

where b is a second constant of integration. The two integrals of Eqs. (1.4.10) that we seek are then

$$u = x^2 + y^2 \tag{1.4.12a}$$

$$v = \psi - \arctan\left(\frac{y}{x}\right) \tag{1.4.12b}$$

The most general solution for ψ is obtained by setting v equal to an arbitrary function H of u:

$$\psi = H(x^2 + y^2) + \arctan\left(\frac{y}{x}\right) \tag{1.4.13}$$

The interpretation of Eq. (1.4.13) is simplified if we write it in polar coordinates:

$$\psi = H(r^2) + \theta \tag{1.4.14}$$

It is now easy to see that the group (1.1.2a, b) transforms the curves of the family (1.4.5): $\psi = c_1$ into one another. If we rewrite Eqs. (1.1.2a, b) in polar coordinates, they become

$$r' = r \qquad\qquad (1.4.15a)$$
$$\theta' = \theta + \lambda \qquad\qquad (1.4.15b)$$

Then

$$\psi(x', y') = H(r'^2) + \theta' = H(r^2) + \theta + \lambda = c_1 + \lambda \quad (1.4.16)$$

Thus the image of the curve labeled by the parameter c_1 is the curve labeled by the parameter $c_1 + \lambda$.∎

1.5 Transformation of Derivatives: the Extended Group

Since the transformations (1.1.1a, b) for fixed λ determine the image C' in the primed plane of any curve C in the unprimed plane, it must be possible to calculate the slope $\dot{y}' \equiv dy'/dx'$ of C' from the slope $\dot{y} \equiv dy/dx$ of C. If $P: (x, y)$ and $Q: (x + dx, y + dy)$ are neighboring points on C, the coordinates of their images $P': (x', y')$ and $Q': (x' + dx', y' + dy')$ on C' are given by

$$x' = X(x, y; \lambda) \qquad\qquad (1.5.1a)$$
$$y' = Y(x, y; \lambda) \qquad\qquad (1.5.1b)$$

and

$$x' + dx' = X(x + dx, y + dy; \lambda) \qquad\qquad (1.5.2a)$$
$$y' + dy' = Y(x + dx, y + dy; \lambda) \qquad\qquad (1.5.2b)$$

To lowest order,

$$dx' = X_x\, dx + X_y\, dy \qquad\qquad (1.5.3a)$$
$$dy' = Y_x\, dx + Y_y\, dy \qquad\qquad (1.5.3b)$$

Thus,

$$\dot{y}' = \frac{Y_x + Y_y \dot{y}}{X_x + X_y \dot{y}} \tag{1.5.4}$$

Equations (1.5.1a, b) and (1.5.4) specify a set of extended transformations of the quantities x, y and $\dot{y}$. Geometrically speaking, x, y and $\dot{y}$ define an infinitesimal line element at the point (x, y) having the slope $\dot{y}$. The set of extended transformations thus carry one such line element into another. Do these extended transformations form a group? When the transformation law for $\dot{y}$ is Eq. (1.5.4), they do,[2] and the group is called the *once-extended group*.

The coefficient of the infinitesimal transformation of the first extended group corresponding to $\dot{y}' = d y'/d x'$ is the derivative $(\partial \dot{y}'/\partial \lambda)_{\lambda=\lambda_0}$. Noting that when $\lambda = \lambda_0, Y_x = X_y = 0$ and $X_x = Y_y = 1$, we find from Eq. (1.5.4) that

$$
\begin{aligned}
\eta_1 &\equiv \left(\frac{\partial \dot{y}'}{\partial \lambda} \right)_{\lambda=\lambda_0} \\
&= \left[\frac{\eta_x + \eta_y \dot{y}}{X_x + X_y \dot{y}} - (Y_x + Y_y \dot{y}) \frac{\xi_x + \xi_y \dot{y}}{(X_x + X_y \dot{y})^2} \right]_{\lambda=\lambda_0} \\
&= (\eta_x + \eta_y \dot{y}) - \dot{y}(\xi_x + \xi_y \dot{y}) \tag{1.5.5a}
\end{aligned}
$$

or

$$\boxed{\eta_1 = \frac{d\eta}{dx} - \dot{y}\frac{d\xi}{dx}} \tag{1.5.5b}$$

Here the two terms in parentheses in the previous line have been written as total directional derivatives in the direction whose slope is $\dot{y}$. Equation (1.5.5) is somewhat easier to remember when written with these total derivatives. The importance of Eq. (1.5.5) is that it is possible to find the coefficient η_1 of the infinitesimal transformation

$$\dot{y}' = \dot{y} + \eta_1(x, y, \dot{y})(\lambda - \lambda_0) \tag{1.5.6}$$

directly from the coefficients ξ and η.

1.6   Transformation of Derivatives (continued)

The formula (1.5.5) for the coefficient η_1 can be derived more easily if we start not with the finite transformations (1.5.1*a*, *b*) but with their infinitesimal form

$$x' = x + \xi(\lambda - \lambda_0) \qquad (1.2.2a)$$
$$y' = y + \eta(\lambda - \lambda_0) \qquad (1.2.2b)$$

Then

$$dx' = dx + d\xi(\lambda - \lambda_0) \qquad (1.6.1a)$$
$$dy' = dy + d\eta(\lambda - \lambda_0) \qquad (1.6.1b)$$

Upon dividing these equations, we find

$$\dot{y}' \equiv \frac{dy'}{dx'} = \left[\dot{y} + \frac{d\eta}{dx}(\lambda - \lambda_0)\right]\left[1 + \frac{d\xi}{dx}(\lambda - \lambda_0)\right]^{-1}$$
$$= \dot{y} + \left(\frac{d\eta}{dx} - \dot{y}\frac{d\xi}{dx}\right)(\lambda - \lambda_0) \qquad (1.6.2)$$

Because the infinitesimals $d\xi$ and $d\eta$ are the differences of ξ and η, respectively, between the points Q and P, the derivatives on the right-hand sides of Eq. (1.6.2) are total directional derivatives in the direction $\dot{y}$. The quantity in parenthesis is the infinitesimal coefficient η_1.

Since the transformations (1.1.1*a*, *b*) for fixed λ determine the image C' in the primed plane of a curve C in the unprimed plane, it must also be possible to calculate the kth derivative $y^{(k)'} \equiv d^k y'/dx'^k$ of C' from the kth derivative $y^{(k)} \equiv d^k y/dx^k$ of C. We can adapt Eq. (1.6.2) to find the infinitesimal coefficient η_k corresponding to $y^{(k)}$ as follows. Since

$$dy^{(k)'} = dy^{(k)} + d\eta_k(\lambda - \lambda_0) \qquad (1.6.3)$$

it follows that

$$y^{(k+1)'} \equiv \frac{dy^{(k)'}}{dx'} = \left[y^{(k+1)} + \frac{d\eta_k}{dx}(\lambda - \lambda_0)\right]\left[1 + \frac{d\xi}{dx}(\lambda - \lambda_0)\right]^{-1}$$
$$= y^{(k+1)} + \left(\frac{d\eta_k}{dx} - y^{(k+1)}\frac{d\xi}{dx}\right)(\lambda - \lambda_0) \qquad (1.6.4)$$

Thus

$$\eta_{k+1} = \frac{d\eta_k}{dx} - y^{(k+1)}\frac{d\xi}{dx} \qquad (1.6.5)$$

The evaluation of the total derivative $d\eta_k/dx$ in Eq. (1.6.5) requires a bit of discussion. One can see inductively that η_k is a function of $x, y, \dot{y}, y^{(2)}, y^{(3)}, \ldots, y^{(k)}$. Therefore,

$$\frac{d\eta_k}{dx} = \frac{\partial \eta_k}{\partial x} + \frac{\partial \eta_k}{\partial y}\dot{y} + \frac{\partial \eta_k}{\partial \dot{y}}y^{(2)}$$
$$+ \frac{\partial \eta_k}{\partial y^{(2)}}y^{(3)} + \ldots + \frac{\partial \eta_k}{\partial y^{(k)}}y^{(k+1)} \qquad (1.6.6)$$

Because of the profusion of terms appearing in these total derivatives, the expressions for the η_k rapidly become very complicated as k increases.

Example: For the rotation group, which we have been using as an example, $\xi = -y$ and $\eta = x$, so that $\eta_1 = 1 + \dot{y}^2$. This can be verified by direct calculation from the polar form (1.4.15) of the group. For, in the primed plane, the angle ϕ' which the tangent to C' at P' makes with the x'-axis is just $\phi + \lambda$, where ϕ is the angle which the tangent to C at P makes with the x-axis. Thus the slope $m' = \tan\phi'$ is given by

$$m' = \tan(\phi + \lambda) = \frac{m + \tan\lambda}{1 - m\tan\lambda} \qquad (1.6.7)$$

where $m = \tan\phi$. Expanding in powers of λ, we find

$$m' = m + \lambda(1 + m^2) + \ldots \qquad (1.6.8)$$

which agrees with the result $\eta_1 = 1 + \dot{y}^2$.∎

1.7   Invariant Differential Equations of the First Order

An invariant of the once-extended group is a function $u(x, y, \dot{y})$ of x, y and $\dot{y}$ whose value at an image point is the same as its value at the source point:

$$u(x', y', \dot{y}') = u(x, y, \dot{y}) \qquad (1.7.1)$$

If we differentiate Eq. (1.7.1) partially with respect to λ and then set $\lambda = \lambda_0$, we obtain the following first-order linear partial differential equation for u

$$\xi u_x + \eta u_y + \eta_1 u_{\dot{y}} = 0 \qquad (1.7.2a)$$

the characteristic equations of which are

$$\frac{dx}{\xi(x, y)} = \frac{dy}{\eta(x, y)} = \frac{d\dot{y}}{\eta_1(x, y, \dot{y})} \qquad (1.7.2b)$$

These equations have two independent integrals and the most general solution for u is an arbitrary function of these two integrals.

A first-order ordinary differential equation is a function $v(x, y, \dot{y})$ of x, y and $\dot{y}$ set equal to zero. If the function v has the property (1.7.2a) when $v = 0$, the differential equation it defines is said to be invariant to the group with the infinitesimal coefficients ξ, η and η_1.

Example: Consider the stretching group $x' = \lambda x$, $y' = \lambda^\beta y$, where β is some fixed number and λ, the group parameter, takes all positive values. Then $\xi = x$, $\eta = \beta y$, $\eta_1 = (\beta - 1)\dot{y}$. Two independent integrals of Eqs. (1.7.2) are y/x^β and $\dot{y}/x^{\beta-1}$. Thus the most general first-order differential equation invariant to this stretching group can be written $\dot{y}/x^{\beta-1} = F(y/x^\beta)$, where F is an arbitrary function.■

Notes

Note 1: Property (1), called the *group property*, says that if $x' = X(x, y; \lambda_1)$, $y' = Y(x, y; \lambda_1)$ and $x'' = X(x', y'; \lambda_2)$, $y'' = Y(x', y'; \lambda_2)$, there is a value of the parameter λ, λ_3, for which $x'' = X(x, y; \lambda_3)$, $y'' = Y(x, y; \lambda_3)$. The value λ_3 is a function $g(\lambda_1, \lambda_2)$ of the values λ_1 and λ_2. In the case of the group (1.1.2) of rotations in a plane, $g(\lambda_1, \lambda_2) = \lambda_1 + \lambda_2$. In general, $g(\lambda_1, \lambda_2) \neq g(\lambda_2, \lambda_1)$, which means that in general the final image point (x'', y'') depends on the order in which two transformations are carried out. In other words, the transformations of the group are not generally *commutative*. Groups like the group (1.1.2) for which the transformations are commutative are called *Abelian* groups.

Although group transformations are not necessarily commutative, they are *associative*: if three transformations (call them 1, 2 and 3) are carried out in succession, it does not matter which pair is carried out first as long as the order 123 is maintained. This means that function g is constrained by the condition $g(\lambda_1, g(\lambda_2, \lambda_3)) = g(g(\lambda_1, \lambda_2), \lambda_3)$.

Note 2: To prove that the set of extended transformations (1.5.1*a, b*) and (1.5.4) form a group we begin by verifying the group property. Let us consider carrying out in succession the transformations $x, y, \dot{y} \rightarrow x', y', \dot{y}'$ that corresponds to parameter λ_1 and $x', y', \dot{y}' \rightarrow x'', y'', \dot{y}''$ that corresponds to parameter λ_2. Denote by P and Q two points on the infinitesimal line element at (x, y). Let the images of P and Q under the transformation with parameter λ_1 be P' and Q' and let the images of P' and Q' under the transformation with parameter λ_2 be P'' and Q''. By the manner in which Eq. (1.5.4) was derived, it must yield $\dot{y}' = \text{slope}(P'Q')$ and $\dot{y}'' = \text{slope}(P''Q'')$. Since the transformations (1.5.1*a, b*) alone form a group, we know that there is a single transformation with parameter λ_3 that carries P into P'' and Q into Q''. Again, Eq. (1.5.4) must yield $\dot{y}'' = \text{slope}(P''Q'')$. Thus the single extended transformation with parameter λ_3 is equivalent to the extended transformations with parameters λ_1 and λ_2 carried out in succession.

In a similar fashion, the remaining group postulates can be verified. Thus the set of extended transformations does indeed form a group, the *once-extended group*.

Problems for Chapter 1

1.1 Do the transformations $y' = F[F^{-1}(y) - \lambda]$, $x' = G[G^{-1}(x) + \lambda]$ form a group? Here $F^{-1}(x)$ is the function inverse to $F(x)$, which means that if $y = F(x)$ then $x = F^{-1}(y)$.

1.2

(a) Use the differential equations (1.2.4) to calculate the directional derivative $df/d\lambda$ of an analytic function $f(x, y)$ along an orbit of a group with infinitesimal coefficients $\xi(x, y)$ and $\eta(x, y)$.

(b) Use the result you have obtained in part (a) to calculate the second derivative $d^2 f/d\lambda^2$.

(c) Now generalize, construct a formula for the nth derivative, and write a Taylor series for $f(x', y')$ in terms of $f(x, y)$ and its derivatives, where the points (x', y') and (x, y) lie on the same orbit and correspond respectively to values λ' and λ of the group parameter.

1.3 Apply the Taylor series you found in part (c) of problem 1.2 to the rotation group whose coefficients are $\xi = -y$ and $\eta = x$ and find the formulas $(1.1.2a, b)$ for the group transformations.

1.4

(a) Find a condition that ξ and η must satisfy in order that the images of the orthogonal trajectories of the orbits are also orthogonal trajectories.

(b) If ξ is a function of y only and η is a function of x only, what are the orbits of the group?

(c) If ξ is a function of x only and η is a function of y only, what are the orbits of the group?

1.5 The coefficients $\mu\xi$ and $\mu\eta$, where μ is a function of x and y, determine the same orbits as the coefficients ξ and η. If we insert $\mu\xi$ in place of ξ and $\mu\eta$ in place of η in the condition of part (a) of problem 4, we find a first-order partial differential equation for μ. If it has a solution μ, then the coefficients $\mu\xi$ and $\mu\eta$ determine a group having the specified orbits which carries the orthogonal trajectories of the orbits into one another. Can you find a very simple way to show that there is always such a group? Hint: consider how families of curves are represented parametrically by Eq. (1.4.1).

1.6 Find the transformation equations for the groups having the following coefficients:

(a) $\xi = 1/x$, $\eta = -1/(2y)$;
(b) $\xi = y$, $\eta = x$;
(c) $\xi = x^2$, $\eta = xy$;
(d) $\xi = x - y$, $\eta = x + y$.

2

First-Order Ordinary Differential Equations

2.1 Lie's Integrating Factor

The general solution of a first-order ordinary differential equation is a one-parameter family of curves (called *integral curves*). If the differential equation is invariant to a group, it takes the same form in the primed variables as in the unprimed variables as explained in section 1.7. This means the family of integral curves must be the same in the primed plane as in the unprimed plane, which is to say that the family of integral curves is an invariant family. As shown in section 1.4, such an invariant family may be represented parametrically by an equation of the form $\psi(x, y) = c$ with the function ψ satisfying the partial differential equation (1.4.9).

The differential equation satisfied by the family (1.4.5) is

$$\psi_x \, dx + \psi_y \, dy = 0 \tag{2.1.1}$$

but an equivalent form, obtained by dividing Eq. (2.1.1) by an arbitrary function $\mu(x, y)$, is

$$M(x, y) \, dx + N(x, y) \, dy = 0 \tag{2.1.2}$$

Since Eq. (2.1.2) is not generally a perfect differential, we need the function $\mu(x, y)$ to integrate it and find the function ψ. The function μ is called an *integrating factor*[1] of Eq. (2.1.2) and because it has been defined as the ratio of Eq. (2.1.1) to Eq. (2.1.2), we must have

$$\psi_x = \mu M \qquad \text{and} \qquad \psi_y = \mu N \qquad (2.1.3)$$

If we substitute Eqs. (2.1.3) into Eq. (1.4.9), which ψ must satisfy, we find that

$$\boxed{\mu = (\xi M + \eta N)^{-1}} \qquad (2.1.4)$$

This is Lie's expression for an integrating factor for a differential equation invariant to the group whose infinitesimal coefficients are ξ and η.

Example: We can use Lie's integrating factor to solve the differential equation

$$y = -\frac{y(y^2 - x)}{x^2} \qquad (2.1.5a)$$

or its equivalent

$$y(y^2 - x)\, dx + x^2\, dy = 0 \qquad (2.1.5b)$$

This differential equation is invariant to the once-extended group

$$y' = \lambda^{1/2} y \qquad (2.1.6a)$$

$$x' = \lambda x \qquad 0 < \lambda < \infty \qquad (2.1.6b)$$

$$\dot{y}' = \lambda^{-1/2} \dot{y} \qquad \text{[see Eq. (1.5.4)]} \qquad (2.1.6c)$$

for, if we imagine Eqs. (2.1.5a, b) written in the primed form and substitute for the primed quantities their values given in Eqs. (2.1.6a–c), we again obtain Eqs. (2.1.5a, b) in the unprimed form. The coefficients ξ and η of the infinitesimal transformations are $\xi = x$ and $\eta = y/2$. According to Lie's theorem, Eq. (2.1.4), an integrating factor is

$$\mu = \left(xy^3 - \frac{x^2 y}{2} \right)^{-1} \qquad (2.1.7)$$

so that

$$\psi_x = \mu M = y(y^2 - x)\left(xy^3 - \frac{x^2y}{2}\right)^{-1}$$

$$= \frac{1}{x} + \frac{1}{(x - 2y^2)} \qquad (2.1.8)$$

which can be integrated to give

$$\psi = \ln x + \ln(x - 2y^2) + F(y) \qquad (2.1.9)$$

The constant of integration, $F(y)$, is an arbitrary function of y since the derivative with respect to x in Eq. (2.1.8) is a partial derivative. If we differentiate Eq. (2.1.9) partially with respect to y and use the second of Eqs. (2.1.3) for ψ_y, we find

$$x^2\left(xy^3 - \frac{x^2y}{2}\right)^{-1} = \mu N = -4y(x - 2y^2)^{-1} + \frac{dF}{dy} \qquad (2.1.10)$$

It follows from Eq. (2.1.10) that $dF/dy = -2/y$ so that $F(y) = -2\ln y + c_2$, where c_2 is a constant of integration. With this value of F, Eq. (1.4.5) can be rearranged to read

$$y = x(2x + c)^{-1/2} \qquad (2.1.11)$$

where $c = \exp(c_1 - c_2)$. As expected this family of curves is carried into itself by the transformations of the group (2.1.6), the image of the curve labeled by the parameter c being labeled by the parameter λc.∎

2.2 The Converse of Lie's Theorem

The converse of Lie's theorem states that if μ is an integrating factor of the differential equation $M(x, y)\,dx + N(x, y)\,dy = 0$ and ξ and η are any two functions satisfying $\mu = (\xi M + \eta N)^{-1}$, then the differential equation is invariant to the group whose infinitesimal transformation has coefficients ξ and η.

To see this, observe that if μ given by Eq. (2.1.4) is an integrating factor, then Eqs. (2.1.3) can be written

$$\psi_x = M(\xi M + \eta N)^{-1} \quad \text{and} \quad \psi_y = N(\xi M + \eta N)^{-1} \tag{2.2.1}$$

Then it follows that

$$\xi \psi_x + \eta \psi_y = 1 \tag{2.2.2}$$

which, according to Eq. (1.4.9), means that the family, $\psi(x, y) = $ *constant*, of integral curves is invariant to the group whose infinitesimal transformation has coefficients ξ and η. Invariance of the family of integral curves means invariance of the differential equation since they are logically identical.

Example: We can use the converse of Lie's theorem to find a second group different from the stretching group of Eq. (2.1.6) to which Eq. (2.1.5) is invariant. Equation (2.1.5) has the integrating factor $\mu = y^{-3}$ which can be discovered by seeking a solution μ that depends only on y to the partial differential equation $(\mu M)_y = (\mu N)_x$. If we arbitrarily choose $\xi = 1$, then Eq. (2.1.4) gives $\eta = y/x$. Using these values of ξ and η, we can solve Eq. (1.2.4) and find $y = c_1 x$, $x = \lambda + c_2$, where c_1 and c_2 are constants of integration. If we choose c_1 to be the ratio y/x for the source point and c_2 to be the x-value of the source point, these last equations are equivalent to the transformations

$$y' = y\left(1 + \frac{\lambda}{x}\right) \tag{2.2.3a}$$

$$x' = x + \lambda \tag{2.2.3b}$$

From these equations it follows that (cf. Eq. (1.5.4))

$$\dot{y}' = \dot{y} + \frac{\lambda(x\dot{y} - y)}{x^2}. \tag{2.2.3c}$$

A straightforward calculation verifies that if we write Eq. (2.1.5a) in the primed form and substitute for the primed variables from Eqs. (2.2.3a, b), we obtain Eq. (2.1.5a) in the unprimed form.

Finally, as required, the group (2.2.3) carries the family (2.1.11) into itself, the image of the curve labeled by parameter c being labeled by parameter $c - 2\lambda$.∎

2.3 Invariant Integral Curves

From the converse of Lie's theorem comes the surprising conclusion that there are infinitely many groups that leave a given first-order differential equation invariant. These can be partitioned according to the integrating factor they correspond to. As we shall see next, all groups (ξ, η) having the same integrating factor μ, given by Eq. (2.1.4), leave the same integral curves of Eq. (2.1.2) invariant.

An invariant curve (an orbit) of a group (ξ, η) is given by the coupled differential equations (1.2.4). If such an orbit is also to satisfy the differential equation (2.1.2), we must then have

$$(\xi M + \eta N)\, d\lambda = 0 \tag{2.3.1}$$

or, since $d\lambda$ is arbitrary,

$$\mu^{-1} = \xi M + \eta N = 0 \tag{2.3.2}$$

Therefore, all groups having the same integrating factor μ leave the same integral curves invariant and these curves can be found by setting $\mu^{-1} = 0$.

Example: In this example, we determine the integral curves of differential equation (2.1.5) that are left invariant by each of the two groups (2.1.6) and (2.2.3) under which the differential equation is invariant. The group (2.1.6) has the integrating factor (2.1.7) and so must leave the integral curves $y = 0$ and $y = (x/2)^{1/2}$ invariant. These curves are those of the family (2.1.11) for which $c = \pm\infty$ and $c = 0$, respectively. These are the only values of c left unchanged by multiplication by λ; thus the integral curves that correspond to them are the only ones that are carried into themselves by the transformations (2.1.6).

The group (2.2.3) belongs to the integrating factor $\mu = y^{-3}$ and thus should leave only the integral curve $y = 0$ invariant. Only $c = \pm\infty$ is left unchanged by subtraction of 2λ, so only the integral curve that corresponds to it, namely $y = 0$, is carried into itself by the transformations (2.2.3).∎

Groups exist that leave any particular integral curve $\psi(x, y) = c_0$ invariant. The most general integrating factor ν of Eq. (2.1.2) has the form

$$\nu = \mu \, F(\psi) \qquad (2.3.3)$$

where μ is some particular integrating factor and F is an arbitrary function of ψ. If μ^{-1} does not already equal zero for the integral curve, we can make $\nu^{-1} = 0$ by choosing F so that $F(c_0) = \infty$.

2.4 Singular Solutions

General as the work of the last two sections may seem, it is nonetheless based on the tacit assumption that at each point of the (x, y)-plane, except for those at which M and N vanish simultaneously, there is but one slope. There are first-order differential equations that do not have the form (2.1.2) for which this is not so. A good example is the differential equation

$$\dot{y}^2 - 2\dot{y} + 4y - 4x = 0 \qquad (2.4.1)$$

for which there are two real slopes $\dot{y}$ at each point (x, y) for which $y < x + 1/4$, one at points for which $y = x + 1/4$, and none at other points. It can be shown by direct substitution that the one-parameter family

$$y = c - (x - c)^2 \qquad (2.4.2)$$

where c is a constant, $-\infty < c < \infty$, satisfies Eq. (2.4.1). Points (x, y) for which $y < x + 1/4$ lie on two curves of the family (2.4.2), points for which $y = x + 1/4$ lie on one curve, and other points on none.

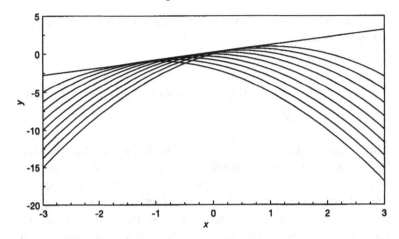

Figure 2.4.1. Curves of the family $y = c - (c - x)^2$ and their envelope $y = x + 1/4$.

Thus the form $\psi(x, y) = c$ where ψ is a single-valued function is inadequate to represent the family (2.4.2) (see figure 2.4.1).

The family (2.4.2) has the *envelope*

$$y = x + \tfrac{1}{4} \tag{2.4.3}$$

The envelope of a family of curves is a curve that is tangent at each of its points to a member of the family. The envelope, which is not itself a member of the family, also satisfies the differential equation of the family $u(x, y, \dot{y}) = 0$. At any point of the envelope (x, y) its slope $\dot{y}$ is the same as that of one of the curves of the family and thus satisfies $u(x, y, \dot{y}) = 0$. Thus Eq. (2.4.3) must also satisfy the differential equation (2.4.1); direct substitution shows that it does. This solution, like any other solution that is not a member of the one-parameter family of integral curves (2.4.2), is called a *singular solution*.

The differential equation (2.4.1) is invariant to the once-extended group

$$x' = x + \lambda$$
$$y' = y + \lambda \qquad -\infty < \lambda < \infty \qquad (2.4.4)$$
$$\dot{y}' = \dot{y}$$

This group carries the curve of the family (2.4.2) labeled by c into the curve labeled by $c + \lambda$, but it carries the envelope (2.4.3) into itself. In fact, any group that leaves the differential equation (2.4.1) invariant must also leave the envelope invariant. For the group permutes the curves of the family among themselves and thus leaves the family as a whole unchanged; therefore it leaves the envelope of the family unchanged. An envelope, if it exists, is thus invariant to *all* the groups that leave the differential equation invariant.

2.5   Change of Variables

Lie proposed a second method of solving first-order differential equations that are invariant to a group, which involves using the group to find new variables $\underline{x}(x, y)$ and $\underline{y}(x, y)$, in terms of which the differential equation becomes separable. (As Cohen points out in his book [Co-11], this method of Lie's was published in 1869, five years before Lie published the formula (2.1.4) for his integrating factor, and so in chronological order this is Lie's first method, not his second.)

Let the new variables $\underline{x}$ and $\underline{y}$ be related to the old variables x and y by the equations

$$\underline{x} = F(x, y) \qquad (2.5.1a)$$
$$\underline{y} = G(x, y) \qquad (2.5.1b)$$

When the point (x, y) is carried into its image (x', y') by a transformation of the group, its new coordinates $(\underline{x}, \underline{y})$ are carried into new image coordinates

$$\underline{x}' = F(X(x, y; \lambda), Y(x, y; \lambda)) \qquad (2.5.2a)$$

$$\underline{y}' = G(X(x, y; \lambda), Y(x, y; \lambda)) \qquad (2.5.2b)$$

We find the coefficients $(\underline{\xi}, \underline{\eta})$ of the infinitesimal transformation in the new variables by differentiating Eqs. (2.5.1a, b) with respect to λ and setting $\lambda = \lambda_0$:

$$\underline{\xi} = \xi F_x + \eta F_y = U\underline{x} \qquad (2.5.3a)$$

$$\underline{\eta} = \xi G_x + \eta G_y = U\underline{y} \qquad (2.5.3b)$$

where U is an abbreviation for the differential operator $\xi\, \partial/\partial x + \eta\, \partial/\partial y$.

Suppose now that we choose the functions F and G so that $\underline{\xi}(\underline{x}, \underline{y}) = 0$ and $\underline{\eta}(\underline{x}, \underline{y}) = 1$. Then $\eta_1(\underline{x}, \underline{y}) = 0$ and Eq. (1.7.2a) becomes

$$\frac{\partial u}{\partial \underline{y}} = 0 \qquad (2.5.4)$$

The quantities $\underline{x}$ and $\underline{\dot{y}}$ are two independent integrals of Eq. (2.5.4)† and thus the most general differential equation invariant to the group (ξ, η) is

$$\underline{\dot{y}} = H(\underline{x}) \qquad (2.5.5)$$

where H is an arbitrary function. Now Eq. (2.5.5) is clearly separable, so if we can solve Eqs. (2.5.3a, b), which now take the form

$$\xi F_x + \eta F_y = 0 \qquad (2.5.6a)$$

$$\xi G_x + \eta G_y = 1 \qquad (2.5.6b)$$

---

† If one or more of the ξ, η and η_1 are zero, as in the case of Eq. (2.5.4), some authors write the characteristic equations (1.7.2b) with zero denominators. In the case of Eq. (2.5.4) the characteristic equations would then be written $d\underline{x}/0 = d\underline{y}/1 = d\underline{\dot{y}}/0$. Strictly speaking zero denominators are not permissible, but the zero denominators are meant only to identify variables that are integrals of the partial differential equation (1.7.2a). When understood in this way, the zero denominators always lead to correct results.

for F and G, we can make a change of variables that will cause our differential equation to become separable.

Example: In this example, we use Lie's second method to solve the differential equation (2.4.1). As noted earlier, the differential equation (2.4.1) is invariant to the translation group (2.4.4) for which $\xi = 1$ and $\eta = 1$. The characteristic equations corresponding to Eqs. (2.5.6a, b) are then, respectively,

$$dx = dy \qquad (2.5.7a)$$

and

$$dx = dy = dG \qquad (2.5.7b)$$

The difference $y - x$ is an integral of Eq. (2.5.7a) so F can be any arbitrary function of $y - x$. Two independent integrals of Eq. (2.5.7b) are $G - y$ and $y - x$ so that G equals y plus an arbitrary function of $y - x$. Thus one satisfactory choice of new variables $\underline{x}$ and $\underline{y}$ is

$$\underline{x} = y - x \qquad (2.5.8a)$$
$$\underline{y} = y \qquad (2.5.8b)$$

From Eqs. (2.5.8a, b) it follows that $y = \underline{y}$, $x = \underline{y} - \underline{x}$, and $\dot{y} = \underline{\dot{y}}/(\underline{\dot{y}} - 1)$ so that upon substitution, Eq. (2.4.1) takes the separated form

$$\underline{\dot{y}} = 1 \pm (1 - 4\underline{x})^{-1/2} \qquad (2.5.9)$$

Integrating Eq. (2.5.9), we find

$$\underline{y} = \underline{x} \pm \tfrac{1}{2}(1 - 4\underline{x})^{1/2} + c \qquad (2.5.10)$$

where c is a constant of integration. Substituting from Eqs. (2.5.8a, b) for $\underline{x}$ and $\underline{y}$, we find, after rearrangement

$$y = -\left[x - \left(c + \tfrac{1}{2}\right)\right]^{2} + \left(c + \tfrac{1}{2}\right) \qquad (2.5.11)$$

which is identical to Eq. (2.4.2).■

Another choice of $\underline{\xi}$ and $\underline{\eta}$ that will cause the variables to separate is $\underline{\xi} = \underline{x}$ and $\underline{\eta} = \overline{0}$. Then $\underline{\eta}_1 = -\underline{\dot{y}}$ and Eqs. (1.7.2) become

$$\frac{\mathrm{d}\underline{x}}{\underline{x}} = \frac{\mathrm{d}\underline{y}}{0} = -\frac{\mathrm{d}\underline{\dot{y}}}{\underline{\dot{y}}} \qquad (2.5.12)$$

The quantities $\underline{x}\underline{\dot{y}}$ and $\underline{y}$ are integrals of Eq. (2.5.12) and thus the most general differential equation invariant to the group $(\underline{\xi}, \underline{\eta})$ is

$$\underline{x}\underline{\dot{y}} = H(\underline{y}) \qquad (2.5.13)$$

which is also separable. So if we can solve Eqs. (2.5.3a, b), which now take the form

$$\xi F_x + \eta F_y = \underline{x} = F \qquad (2.5.14a)$$
$$\xi G_x + \eta G_y = 0 \qquad (2.5.14b)$$

for F and G, we can make another change of variables that will also cause our differential equation to separate.

Theorem 2.5.1. If $X(x, y; \lambda) = \lambda x$, i.e. if the transformation of x by the group is a stretching, then $\xi = x$ and we can satisfy Eq. (2.5.14a) by choosing $F = x$. The function G is a group invariant (see Eq. (1.3.2)). *Thus for groups in which the transformation of x is a stretching, introducing a group invariant as a new variable in place of y and leaving the variable x unchanged results in a new differential equation that is separable.*

Corollary 2.5.2. Similarly if $X(x, y; \lambda) = x + \lambda$, i.e. if the transformation of x by the group is a translation, then $\xi = 1$ and we can again satisfy Eq. (2.5.14a) by choosing $F = \mathrm{e}^x$. The function G is again a group invariant (see Eq. (1.3.2)) so that introducing a group invariant as a new dependent variable in place of y and using e^x as the other new variable also results in a new differential equation that is separable. After the new variables are separated we can again use x in place of e^x. *Thus for groups in which the transformation of x is a translation, introducing a group invariant as a new variable in place of y and leaving the variable x unchanged results in a new differential equation that is separable.*

These theorems are convenient because stretching and translation groups are quite common in practice. The theorems apply as well to the cases in which the variable y is stretched or translated.

★2.6   Tabulation of Differential Equations

The theory outlined in section 1.7 shows how to find the most general differential equation that is invariant to a given group. To get an explicit representation of this differential equation, one must find explicit representations of two integrals of Eqs. (1.7.2). One of these two explicit integrals also provides a solution (at least in the form of an integral) to Eqs. (1.2.4). Thus we can find the finite form of the group (again, at least in the form of an integral).

Example: Let us consider groups for which ξ is a function of x and η is a function of y, i.e. groups for which

$$\xi = P(x) \qquad \text{and} \qquad \eta = Q(y) \qquad (2.6.1)$$

According to Eq. (1.5.5),

$$\eta_1 = \dot{y}\left(\frac{\mathrm{d}Q}{\mathrm{d}y} - \frac{\mathrm{d}P}{\mathrm{d}x}\right) \qquad (2.6.2)$$

Two integrals of Eqs. (1.7.2) are then

$$\int \frac{\mathrm{d}x}{P} - \int \frac{\mathrm{d}y}{Q} \qquad \text{and} \qquad \frac{\dot{y}P}{Q} \qquad (2.6.3)$$

The most general differential equation invariant to the group (2.6.1) is thus

$$\dot{y} = \frac{Q(y)}{P(x)} F\left(\int \frac{\mathrm{d}x}{P} - \int \frac{\mathrm{d}y}{Q}\right) \qquad (2.6.4)$$

where F is an arbitrary function. Integration of Eqs. (1.2.4) yields the finite form of the group, namely,

$$\int_{x}^{x'} \frac{dt}{P(t)} = \lambda \qquad (2.6.5a)$$

$$\int_{y}^{y'} \frac{dt}{Q(t)} = \lambda \qquad (2.6.5b)$$

The following are some interesting special cases of these formulas:

Case 1: $P = a$, $Q = b$, where a and b are constants. Then

$$\dot{y} = \frac{b}{a}F\left(\frac{x}{a} - \frac{y}{b}\right) = G(cx - y) \qquad \text{where} \quad c = \frac{b}{a} \quad (2.6.6a)$$

$$x' = x + a\lambda = x + \mu \qquad \text{where} \quad \mu = a\lambda \qquad (2.6.6b)$$

$$y' = y + b\lambda = y + c\mu \qquad (2.6.6c)$$

and F and G are arbitrary functions. The group is a pure translation group.

Case 2: $P = ax$, $Q = by$, where a and b are constants. Then

$$\dot{y} = \frac{by}{ax}F\left(\frac{\ln x}{a} - \frac{\ln y}{b}\right) = \frac{y}{x}G\left(\frac{y}{x^c}\right) \qquad \text{where} \quad c = \frac{b}{a}$$

$$\qquad (2.6.7a)$$

$$x' = xe^{a\lambda} = \mu x \qquad \text{where} \quad \mu = e^{a\lambda} \qquad (2.6.7b)$$

$$y' = ye^{b\lambda} = \mu^c y \qquad (2.6.7c)$$

This group is a pure stretching group.

Case 3: $P = a/x$, $Q = by$, where a and b are constants. Then

$$\dot{y} = xyG[y\exp(-cx^2)] \qquad \text{where} \quad c = \frac{b}{2a} \qquad (2.6.8a)$$

$$x' = (x^2 + \mu)^{1/2} \qquad \text{where} \quad \mu = 2a\lambda \qquad (2.6.8b)$$

$$y' = e^{c\mu}y \qquad (2.6.8c)$$

I have added this last case to show what complexity may follow from a comparatively simple choice of ξ and η.∎

There are other choices of ξ and η for which Eqs. (1.7.2) and (1.2.4) are integrable, and it is possible to build up tables of differential equations and groups that leave them invariant. One such table may be found in Cohen's book [Co-11].

Notes

Note 1:   I note here some useful and well-known facts about integrating factors. The differential equation (2.1.2) has infinitely many integrating factors corresponding to the infinitely many ways that its family of integral curves may be parametrized in the form (1.4.5). Each integrating factor μ satisfies the partial differential equation $(\mu M)_y = (\mu N)_x$. If μ and ν are two integrating factors and one is not a constant multiple of the other, then the family of integral curves of the differential equation is given by $\nu/\mu = c$, where c is a parameter labeling the integral curves. Finally, if μ is an integrating factor and the integral curves are parametrized as in Eq. (1.4.5), then: (1) $\nu = \mu F(\psi)$, where F is an arbitrary function, is another integrating factor; and (2) every other integrating factor ν has the form $\mu F(\psi)$. These results can be found in any standard text on differential equations, for example, [Fo-33].

Problems for Chapter 2

2.1   Find by inspection a group under which the differential equation $2x^4 y\dot{y} + 4x^3 y^2 + 2x = 0$ is invariant, calculate Lie's integrating factor and find the general solution of the differential equation.

2.2   Find a group which leaves the differential equation $\dot{y} = xy(1 + \ln y + x^2)$ invariant, and using Lie's integrating factor find the general solution.

2.3   The linear, first-order, inhomogeneous differential equation $\dot{y} + P(x)\, y = Q(x)$ has the well-known integrating factor $\exp(\int_0^x P(z)\, dz)$. (Here z is a dummy variable of integration. The choice of the lower limit of integration will not affect the results and I have arbitrarily chosen it to be zero here for definiteness.) Use the converse of Lie's theorem (section 2.2) to find the transformation equations of two different groups to which this differential equation is invariant.

2.4 Find a group under which the differential equation $\dot{y} = ax + by + c$ is invariant, calculate Lie's integrating factor and find the integral curves of the differential equation.

2.5 Verify that the differential equation $\dot{y} = (\ln y - e^x) \exp(x + e^x)$ is invariant to the group $x' = \ln(e^x + \lambda)$, $y' = \lambda y$. Use Lie's second method (section 2.5) to determine a change of variables that separates the variables and find the form the differential equation takes in the new variables.

2.6 Prove that the differential equation $u(x, y, \dot{y}) = 0$ satisfies Eq. (1.7.2a) if its family of integral curves $\psi(x, y) = c$ satisfies Eq. (1.4.9). Thus prove that the differential equation is invariant if its family of integral curves is an invariant family. Can your proof be inverted to prove the converse, namely, that if the differential equation is invariant then its family of integral curves is invariant?

2.7 Goursat gives the integrating factor $v = x^p y^q$ for the differential equation $y(a + \alpha x^m y^n) \, dx + x(b + \beta x^m y^n) \, dy = 0$, where p and q are determined by the simultaneous equations $bp - aq = a - b$ and $\beta p - \alpha q = \alpha(n + 1) - \beta(m + 1)$ [Da-62]. Find a group to which Goursat's differential equation is invariant, calculate Lie's integrating factor μ and, if it is not a multiple of Goursat's, use it to find the general solution by setting $\mu/v = c$, an arbitrary constant. (Remember that the ratio of two integrating factors which are not constant multiples of each other gives the general solution when set equal to an arbitrary constant.)

2.8 Davis [Da-62] states that the function $\mu = (x^2 + y^2)^{-1}$ is an integrating factor for the equation $(y + xF) \, dx - (x - yF) \, dy = 0$ where $F = F(x^2 + y^2)$. Prove this by calculating the most general differential equation of the first order, invariant to the rotation group $\xi = -y$, $\eta = x$, showing that Lie's integrating factor for this most general differential equation is $\mu = (x^2 + y^2)^{-1}$, and finally showing this most general differential equation is identical with Davis's equation.

2.9 Chrystal's differential equation, $\dot{y}^2 + Ax\dot{y} + By + Cx^2 = 0$, has a singular solution for certain values of the coefficients A, B and C [Da-62]. Find a group to which Chrystal's equation is invariant, use the group to determine this singular solution, and find a *necessary* condition that the coefficients A, B and C must obey for a singular solution to exist.

2.10 A curve C_2 is said to be *parallel* to a curve C_1 if the distance from C_1 along its normal to the curve C_2 is the same for all points of C_1.

(a) Prove that any normal to the curve C_1 is also normal to the curve C_2.

(b) Imagine the (x, y)-plane filled with the one-parameter family F of curves parallel to C_1. Denote by C_δ the curve parallel to C_1 at a distance δ from C_1. Define the set of transformations T_λ of points P: (x, y) into images P': (x', y') in the following way. Through the point P passes a curve C_μ of the family F. Take as the image point P' of P under the transformation T_λ the point (x', y') lying at a distance λ from P along the normal to C_μ. Using the result of part (a), show that the set of transformations $\{T_\lambda\}$, $-\infty < \lambda < \infty$, form a one-parameter group G.

(c) If the differential equation of the family F is $M(x, y)\,dx + N(x, y)\,dy = 0$, show that the coefficients ξ and η of the infinitesimal transformation of the group G are $\xi = M/(M^2 + N^2)^{1/2}$ and $\eta = N/(M^2 + N^2)^{1/2}$. Show, therefore, that Lie's integrating factor for this differential equation is $\mu = (M^2 + N^2)^{-1/2}$. (According to Cohen [Co-11, p. 69], Lie derived this integrating factor for the differential equation of a family of parallel curves by purely geometric considerations, apparently without the use of his theory of groups.)

(d) The converse of part (c) is also true, namely, if $\mu = (M^2 + N^2)^{-1/2}$ is an integrating factor of the differential equation $M(x, y)\,dx + N(x, y)\,dy = 0$, then the integral curves of this differential equation are parallel to one another. Can you prove this result, too?

2.11

(a) Show that $\mu = e^x$ is an integrating factor for the differential equation

$$M(x, y)\,dx + N(x, y)\,dy = (3x^2y + y^3 + 6xy)\,dx$$
$$+ 3(x^2 + y^2)\,dy = 0$$

Hint: consider the partial differential equation $(\mu M)_y = (\mu N)_x$.

(b) Find the y-coefficient η of the infinitesimal transformation of a group G that leaves the differential equation invariant and for which the x-coefficient $\xi = 0$.

(c) Determine the transformation equations of the group G.

2.12 The integral curves of the Clairaut equation $y = x\dot{y} + (1 + \dot{y}^2)^{1/2}$ form a family of straight lines that define an envelope.

(a) Show that this differential equation is invariant to the rotation group $\xi = y$, $\eta = -x$.

(b) Use this group to show that the envelope is the circle of unit radius centered on the origin.

2.13 The integral curves of the differential equation

$$\dot{y}^2 - 8x^3\dot{y} + 16x^2y = 0$$

define an envelope. Find a group that leaves the differential equation invariant. Determine from the group the equation of the envelope.

2.14

(a) Use the fact that the differential equation

$$(2x^2y - x^3 - y)\,dx + x\,dy = 0$$

is linear to find an integrating factor for it (see problem 2.3).

(b) Use the results of part (a) to find a substitution for y that separates the variables in the equation.

(c) Use this new variable to determine the coefficients ξ and η of the infinitesimal transformation of a group G under which the equation is invariant. [Hint: consider Eqs. (2.5.6).]

(d) Determine the coefficients ξ and η directly from the differential equation and thereby check the result of part (c).

2.15

(a) If the transformation equations (1.1.1a, b) of a group G can be algebraically manipulated into the form

$$u(x', y') = u(x, y) \tag{1a}$$

$$v(x', y') = v(x, y) + \lambda \tag{1b}$$

show that introducing u and v as new variables in any first-order differential equation invariant under G yields a differential equation in u and v in which these variables separate.

(b) Find a group that leaves the differential equation

$$y\dot{y} + \exp(-y^2)(y^2 - \ln x)^{1/2} = 0 \tag{2}$$

invariant.

(c) Use the result of part (a) to determine the new variables u and v and find the transformed differential equation in the variables u and v.

2.16   The substitution $y = u^{1/(1-n)}$ linearizes the Bernoulli equation

$$\frac{dy}{dx} + P(x)\, y = Q(x)\, y^n$$

(a) Using the integrating factor for linear equations (see problem 2.3), find a substitution for u that separates the variables. Thus find a substitution for y that separates the variables in the Bernoulli equation.

(b) Use Eqs. (2.5.6) to calculate ξ and η for a group that leaves the Bernoulli equation invariant.

(c) Determine relations that the coefficients ξ and η must obey directly from the Bernoulli equation and use them to check your results from part (b).

2.17 Find a group which leaves invariant both the integral curve $y = c_0 e^x$ of the elementary differential equation $\dot{y} = y$ and the differential equation itself.

3

Second-Order Ordinary Differential Equations

3.1 Invariant Differential Equations of the Second Order

An invariant of the twice-extended group is a function $w(x, y, \dot{y}, \ddot{y})$ of $x, y, \dot{y}$ and $\ddot{y}$ whose value at an image point is the same as its value at the source point:

$$w(x', y', \dot{y}', \ddot{y}') = w(x, y, \dot{y}, \ddot{y}) \qquad (3.1.1)$$

If we differentiate Eq. (3.1.1) partially with respect to λ and set $\lambda = \lambda_0$, we obtain the following first-order linear partial differential equation for w

$$\xi w_x + \eta w_y + \eta_1 w_{\dot{y}} + \eta_2 w_{\ddot{y}} = 0 \qquad (3.1.2a)$$

the characteristic equations of which are

$$\frac{dx}{\xi(x, y)} = \frac{dy}{\eta(x, y)} = \frac{d\dot{y}}{\eta_1(x, y, \dot{y})} = \frac{d\ddot{y}}{\eta_2(x, y, \dot{y}, \ddot{y})} \qquad (3.1.2b)$$

These equations have three independent integrals and the most general solution for w is an arbitrary function of these three integrals.

A second-order ordinary differential equation can be written in the form

$$w(x, y, \dot{y}, \ddot{y}) = 0 \qquad (3.1.3a)$$

If the function w has the property (3.1.2a) when $w = 0$, the differential equation is said to be invariant to the extended group with the infinitesimal coefficients ξ, η, η_1 and η_2.

Differential equations of the second order can always be written as a pair of coupled differential equations of the first order by setting $u = \dot{y}$. Then the differential equation $w(x, y, \dot{y}, \ddot{y}) = 0$ becomes the coupled pair

$$\dot{y} = u \tag{3.1.3b}$$

$$w(x, y, u, \dot{u}) = 0 \tag{3.1.3c}$$

These two equations must be solved *simultaneously*. If the second-order differential equation is invariant to a group (ξ, η), Lie has shown how to reduce the problem of solving it to that of solving two first-order differential equations one at a time *successively*. Lie's way of doing this is explained in the next section.

3.2 Lie's Reduction Theorem

The equations (3.1.3b, c) determine a two-parameter family of curves in the three-dimensional space whose coordinates are x, y and u. (Consider the direction field of infinitesimal vectors dx, $dy = \dot{y}\,dx$, $du = \dot{u}\,dx$ determined by solving Eqs. (3.1.3b, c) for $\dot{y}$ and $\dot{u}$. These vectors define curves in space that can be labeled with their intersection with some surface; thus the curves comprise a two-parameter family.)

When the differential equations (3.1.3b, c) are invariant to the once-extended group (ξ, η, η_1), the transformations of the group carry each of these curves into other curves of the family. Thus each curve belongs to a one-parameter subfamily, the curves of which map into one another under the one-parameter infinitude of transformations of the group. Each such one-parameter family of curves defines a surface in (x, y, u)-space, and from the manner of its definition we see that each such surface maps into itself under the transformations of the group.

These invariant surfaces form a one-parameter family which we denote by the equation

$$\phi(x, y, u, c) = 0 \qquad (3.2.1)$$

The invariance of each individual surface of this family means that

$$0 = \phi(x', y', u', c) = \phi(X(x, y; \lambda), Y(x, y; \lambda), U(x, y, u; \lambda), c) \qquad (3.2.2)$$

If we differentiate Eq. (3.2.2) with respect to λ and set $\lambda = \lambda_0$, we find

$$\xi\phi_x + \eta\phi_y + \eta_1\phi_u = 0 \qquad (3.2.3a)$$

the characteristic equations of which are

$$\frac{dx}{\xi(x, y)} = \frac{dy}{\eta(x, y)} = \frac{du}{\eta_1(x, y, u)} \qquad (3.2.3b)$$

If $p(x, y)$ and $q(x, y, u)$ are two integrals of Eqs. (3.2.3b), the most general solution for ϕ is an arbitrary function G of p and q. Thus the family (3.2.1) can be represented by the equation

$$G(p, q, c) = 0 \qquad (3.2.4)$$

The function $p(x, y)$, being an integral of the first pair of Eqs. (3.2.3b), is a *group invariant*. The function $q(x, y, u) = q(x, y, \dot{y})$, which is an invariant of the once-extended group, is called a *first differential invariant.*

Equation (3.2.4) represents a one-parameter family of curves in the (p, q)-plane and, as we have already seen, a one-parameter family of curves is equivalent to a first-order differential equation. What we have proved so far is *Lie's reduction theorem*:

If we adopt the invariant p and first differential invariant q as new variables, the second-order differential equation $w(x, y, \dot{y}, \ddot{y}) = 0$ will reduce to a first-order differential equation in p and q.

If the (p, q) differential equation, henceforth called the *associated differential equation*, can be solved explicitly, we can obtain an explicit representation of the function G in Eq. (3.2.4). Since the first differential invariant q involves the variable $\dot{y}$, Eq. (3.2.4) is, in fact, a first-order differential equation for y in terms of x. Furthermore, since both p and q are group invariants, Eq. (3.2.4) is invariant to the group (ξ, η). Therefore, we can apply to it the techniques developed in the last chapter, i.e. separation of variables or construction of an integrating factor.

It is worth noting here that if the transformation functions X and Y in Eqs. (1.1.1) are explicitly known, it is possible to calculate the invariant p by algebraic manipulation and the first differential invariant q by differentiation and algebraic manipulation. This is proved in Appendix D.

Example: The Emden–Fowler equation

$$\ddot{y} + \frac{2\dot{y}}{x} + y^n = 0 \qquad (3.2.5)$$

arises in the study of the equilibrium mass distribution in a gas cloud held together by gravitation; the exponent n is related to the adiabatic exponent γ of the gas by $\gamma = (n + 1)/n$. Equation (3.2.5) is invariant to the twice-extended stretching group

$$x' = \lambda x \qquad (3.2.6a)$$
$$y' = \lambda^\beta y \qquad (3.2.6b)$$
$$\dot{y}' = \lambda^{\beta-1} \dot{y} \qquad (3.2.6c)$$
$$\ddot{y}' = \lambda^{\beta-2} \ddot{y} \qquad (3.2.6d)$$

where $\beta = 2/(1 - n)$. If we write Eq. (3.2.5) in the primed form and substitute from Eqs. (3.2.6) for the primed variables, we again obtain Eq. (3.2.5) in the unprimed variables.

An invariant p and a first differential invariant q are

$$p = \frac{y}{x^\beta} \qquad (3.2.7a)$$

$$q = \frac{\dot{y}}{x^{\beta-1}} \qquad (3.2.7b)$$

The choice of these invariants is not unique, and any function of p could be used as an invariant and any function of p and q could be used as a first differential invariant. Let us now calculate dp/dx and dq/dx:

$$\frac{dp}{dx} = \frac{\dot{y}}{x^\beta} - \frac{\beta y}{x^{\beta+1}} = \frac{q - \beta p}{x} \qquad (3.2.8a)$$

$$\frac{dq}{dx} = \frac{\ddot{y}}{x^{\beta-1}} - (\beta - 1)\frac{\dot{y}}{x^\beta} = -\left[(\beta+1)q + p^{\beta-2/\beta}\right]/x \qquad (3.2.8b)$$

Here we have replaced $\ddot{y}$ by its value in terms of x, y and $\dot{y}$ obtained from the Emden–Fowler equation (3.2.5). Dividing Eq. (3.2.8b) by Eq. (3.2.8a) we obtain

$$\frac{dq}{dp} = \frac{-\left[(\beta+1)q + p^{\beta-2/\beta}\right]}{q - \beta p} \qquad (3.2.9)$$

which is a first-order differential equation, as expected.

When $n = 5\,(\beta = -1/2)$, Eq. (3.2.9) can be integrated directly to give

$$3q^2 + 3qp + p^6 = a \qquad (3.2.10)$$

where a is a constant of integration. In terms of x, y and $\dot{y}$, this last equation can be written

$$3x^3\dot{y}^2 + 3x^2 y\dot{y} + x^3 y^6 = a \qquad (3.2.11)$$

Equation (3.2.11) is invariant to the group (3.2.6), as it must be. Because this group is a stretching group, we can use theorem 2.5.1 which states that the introduction of a group invariant as a new dependent variable will cause Eq. (3.2.11) to become separable. For this purpose, the invariant $s = p^2 = xy^2$ is convenient. The most interesting solutions of Eq. (3.2.11) are those for which $a = 0$ because y, being the gravitational potential, must be finite at the origin and have zero derivative there. Then Eq. (3.2.11) becomes

$$\frac{dx}{x} = \frac{\sqrt{3}}{2}\left(\frac{3}{4} - s^2\right)^{-1/2}\frac{ds}{s} \qquad (3.2.12)$$

which can be integrated with the help of the substitution $s = (\sqrt{3}/2)\sin\theta$ to give

$$s = \frac{3bx}{x^2 + 3b^2} \tag{3.2.13a}$$

where b is a constant of integration, or

$$y = \left(\frac{3b}{x^2 + 3b^2}\right)^{1/2} \tag{3.2.13b}$$

This solution is well known [Da-62].■

3.3 Stretching Groups

When the differential equations (3.1.3a–c) are invariant under a stretching group, the group can be used to determine the asymptotic form of certain solutions that are important in applications. Knowing their asymptotic behavior greatly simplifies determining the solutions. Since stretching groups occur frequently in practice, a detailed discussion is warranted.

The most general stretching group in two variables x and y is

$$x' = \lambda x \tag{3.3.1a}$$
$$y' = \lambda^\beta y \qquad 0 < \lambda < \infty \tag{3.3.1b}$$

where β is a constant. No generality is lost by choosing the exponent of λ in Eq. (3.3.1a) to be 1. Then

$$\xi = x \tag{3.3.2a}$$
$$\eta = \beta y \tag{3.3.2b}$$
$$\eta_1 = (\beta - 1)\dot{y} \tag{3.3.2c}$$
$$\eta_2 = (\beta - 2)\ddot{y} \tag{3.3.2d}$$

It follows therefore that the most general second-order differential equation invariant to the group (3.3.1) is an arbitrary function of three independent integrals of the characteristic equations (3.1.2*b*), which now take the form

$$\frac{dx}{x} = \frac{dy}{\beta y} = \frac{d\dot{y}}{(\beta - 1)\dot{y}} = \frac{d\ddot{y}}{(\beta - 2)\ddot{y}} \qquad (3.3.3)$$

The functions y/x^β, $\dot{y}/x^{\beta-1}$ and $\ddot{y}/x^{\beta-2}$ are three such integrals, so the most general second-order differential equation invariant to the group (3.3.1) must have the form

$$\phi\left(\frac{y}{x^\beta}, \frac{\dot{y}}{x^{\beta-1}}, \frac{\ddot{y}}{x^{\beta-2}}\right) = 0 \qquad (3.3.4)$$

where ϕ can be any function.

From Eq. (3.3.4) it follows that there are real power-law solutions of the form $y = Ax^\beta$ when A is a real constant that satisfies the algebraic equation

$$\phi(A, \beta A, \beta(\beta - 1)A) = 0 \qquad (3.3.5)$$

Under circumstances outlined below, one of them gives the asymptotic behavior of positive solutions on the half-line $x \geq 0$ that vanish at infinity.

The value of y at $x = 0$ specifies one of these solutions. (The determination of such a solution is called a *two-point boundary-value problem*. Here the boundary values at the two points $x = 0$ and $x = \infty$ are $y(0) = a$, a constant and $y(\infty) = 0$.) All of these solutions are images of one another under the group (3.3.1). Since any image of the point $(0, y_1(0))$ is $(0, \lambda^\beta y_1(0))$, we can always find a value of λ for which $\lambda^\beta y_1(0) = y_2(0)$. Furthermore, since the point $x = \infty$ transforms into the point $x' = \infty$, we see that $y_2(\infty) = \lambda^\beta y_1(\infty) = 0$. Hence the solution $y_2(x)$ obeying the boundary conditions $y_2(0) = a_2$ and $y_2(\infty) = 0$ is an image of the solution $y_1(x)$ obeying the boundary conditions $y_1(0) = a_1$ and $y_1(\infty) = 0$, the value of λ corresponding to the transformation being $(a_2/a_1)^{1/\beta}$.

Because $y_2(x)$ and $y_1(x)$ are images of one another,

$$\frac{y_2(x')}{x'^\beta} = \frac{y_1(x)}{x^\beta} \qquad (3.3.6)$$

where $x' = \lambda x$. Thus the limit of $y(x)/x^\beta$ as $x \to \infty$ is the same for all members of the family. Call it B. So far we know nothing about B: its value can be infinite, finite or zero; and we can go no further using purely group-theoretic arguments. But if we add some additional hypotheses, we can determine the value of B.

For some differential equations for which $\beta < 0$ it happens that the solutions $y(x)$ that vanish at infinity are *ordered* according to their values at $x = 0$. This last statement means that if $y_1(0) > y_2(0) > 0$, then $y_1(x) \geq y_2(x) \geq 0$ for all x. If the solutions are ordered, then the power-law solution $y_* = A_* x^\beta$, where A_* is now the smallest positive root of Eq. (3.3.5), is greater than each of them since $y_*(0) = \infty$ and $y_*(\infty) = 0$ (remember that now $\beta < 0$). Since the solutions of the family are bounded by y_* from above and increase monotonically for fixed x as their initial value $y(0) = a$ increases, they must have a limit as $a \to \infty$. This limit, being arbitrarily close to solutions of the family, must also be a solution. Furthermore, it must be invariant to the group (3.3.1) since the group carries the family of solutions into itself. The limit curve must therefore be of the form $y_{\lim} = Cx^\beta$, $C > 0$, since y/x^β is the group invariant. Because $Cx^\beta = y_{\lim} \geq y(x)$, C must equal A_* since A_* is the smallest positive root of Eq. (3.3.5). Thus the power-law solution $y_* = A_* x^\beta$ is the limit of the family of solutions as $a \to \infty$ and is approached uniformly from below.

Now we are poised for the final step of the proof, which is to show that the limit $B = A_*$. Consider a particular solution S belonging to a fixed value of a. Let us focus our attention on a particular set of images (x', y') of the points (x, y) of S, namely, those for which $\lambda = c/x$, where c is a constant:

$$x' = c \tag{3.3.7a}$$

$$y' = \left(\frac{c}{x}\right)^\beta y \tag{3.3.7b}$$

These images all lie on a single vertical line at $x' = c$. The image point (x', y') also lies on the curve belonging to the initial value

$a' = (c/x)^\beta a$. Thus as $x \to \infty$, $a' \to \infty$ (remember $\beta < 0$). This means as the point (x, y) moves out along the solution S toward $x = \infty$, the image point (x', y') moves steadily upwards along the vertical line $x' = c$. Then, since $y/x^\beta = y'/x'^\beta$,

$$B = \lim\left(\frac{y}{x^\beta}\right) \qquad \text{as} \quad x \to \infty$$

$$= \lim\left(\frac{y'}{x'^\beta}\right) \qquad \text{as} \quad a' \to \infty = A_* \qquad (3.3.8)$$

which completes the proof.

Example: The differential equation

$$4\ddot{y} + 9x\dot{y}^{5/3} = 0 \qquad (3.3.9)$$

arises in the study of heat transfer in superfluid helium. It is invariant to the group (3.3.1) with $\beta = -2$. Substitution of the form $y = Ax^{-2}$ into the differential equation yields $A = 4/(3\sqrt{3})$ so that $y_* = (4/[3\sqrt{3}])x^{-2}$.

To show that the positive solutions for which $y(\infty) = 0$ are ordered, we set $u = \delta y$, the difference between two infinitesimally close neighboring solutions. To first order, the difference u obeys the differential equation

$$4\ddot{u} + 15x\dot{y}^{2/3}\dot{u} = 0 \qquad (3.3.10)$$

In this equation, the quantity $15x\dot{y}^{2/3}$ is treated as a known function of x. A single integration of Eq. (3.3.10) shows that $\dot{u}$ always has one sign. If $u(0) > 0$ and $u(\infty) = 0$, then $\dot{u} < 0$. Thus $u(x) \geq 0$ for all x, for if it were less than zero anywhere, somewhere further on it would have to climb with a positive slope in order to reach $u(\infty) = 0$. This is sufficient to prove the ordering of the positive solutions for which $y(\infty) = 0$.

Equation (3.3.9) has been picked as an example because it is explicitly solvable. A short calculation shows that

$$y = \left(\frac{8}{3\sqrt{3}a^2}\right)[1 - x(x^2 + a^2)^{-1/2}] \qquad (3.3.11)$$

where a is an arbitrary constant, is the solution that vanishes at $x = \infty$. If we expand the right-hand side of Eq. (3.3.11) in reciprocal powers of x, we find that for large x the leading term is $(4/[3\sqrt{3}])x^{-2}$, as anticipated. Note that this asymptotic limit is the same no matter what the value of $y(0)$, as was already evident from the general proof given earlier.■

Example: The Thomas–Fermi equation

$$\ddot{y} = x^{-1/2}y^{3/2} \qquad (3.3.12)$$

arises in atomic physics, where the solution that is sought obeys the two-point boundary conditions $y(0) = 1$ and $y(\infty) = 0$. The differential equation is invariant to the group (3.3.1) with $\beta = -3$. Substituting $y = Ax^{-3}$ into the differential equation yields $A = 144$. If we can prove that the positive solutions for which $y(\infty) = 0$ are ordered, then $y = 144x^{-3}$ is shown to be their asymptotic form.

The differential equation for the infinitesimal difference $\delta y = u$ is

$$\ddot{u} = \tfrac{3}{2}x^{-1/2}y^{1/2}u \qquad (3.3.13)$$

where $u(\infty) = 0$. We wish to show that if $u(0) > 0$, then $u(x) \geq 0$ for all x. We do this by *reductio ad absurdum*. If $u(x)$ were negative anywhere, it would have to have a negative minimum somewhere since it starts out positive and is zero at infinity. But at a negative minimum, $u < 0$ and $\ddot{u} \geq 0$. Since $(3/2)x^{-1/2}y^{1/2} > 0$, these conditions violate Eq. (3.3.13). Thus the assumption that $u(x)$ is negative leads to a contradiction. This is enough to prove the ordering.

If we choose $p = x^3y$ and $q = x^4\dot{y}$ as an invariant and a first differential invariant, we find

$$x\frac{dq}{dx} = 4q + p^{3/2} \qquad (3.3.14a)$$

$$x\frac{dp}{dx} = 3p + q \qquad (3.3.14b)$$

$$\frac{dq}{dp} = \frac{4q + p^{3/2}}{3p + q} \qquad (3.3.14c)$$

The associated differential equation (3.3.14c) cannot be solved explicitly, so how shall we proceed? This is an important question because in many applications of Lie's reduction theorem the associated differential equation cannot be integrated in closed form. We must then turn to numerical integration; but it is not yet clear which among the infinitude of integral curves of the associated differential equation is the one to calculate. The analysis given below suggests a way to determine this integral curve. The method employed applies to many applications including the one described below.

The general solution of the first-order differential equation (3.3.14c), like that of any first-order differential equation, is a one-parameter family of curves $\phi(p, q) = c$. Only one curve of this family corresponds to the family $y(x)$ of positive solutions of the Thomas–Fermi equation (3.3.13) for which $y(\infty) = 0$; for any one of these curves $y(x)$ is uniquely determined by its value $y(0)$ at $x = 0$. An image $y'(x')$ of $y(x)$ has the value $\lambda^{-3}y(0)$ at $x' = 0$ and also vanishes at $x' = \infty$. Thus all the positive solutions $y(x)$ that have finite values at the origin and vanish at infinity are images of one another under transformations of the group (3.3.1). Now since $x'^3 y' = x^3 y$ and $x'^4 \dot{y}' = x^4 \dot{y}$, any image $y'(x')$ of $y(x)$ determines the same locus in the (p, q)-plane as the curve $y(x)$ itself.

The desired curve in the (p, q)-plane can be determined by studying the direction field of the associated differential equation (3.3.14c). (The direction field, the reader may recall, is obtained by drawing a short line segment having the slope dq/dp given by the differential equation at each point (p, q) of the (p, q)-plane.) Since we expect $p > 0$ (since x and y are > 0) and $q < 0$ (since $\dot{y} < 0$), we concentrate on the fourth quadrant, shown in figure 3.3.1.

The curves on which $dq/dp = 0$ or $dq/dp = \pm\infty$ divide the plane into regions in which the algebraic sign of dq/dp is constant. The locus of zero slope C_0 is the curve $4q + p^{3/2} = 0$ and the locus of infinite slope C_∞ is the line $3p + q = 0$. The intersections of these loci are the singular points of the differential equation. They are a node at the origin $O: (0, 0)$ and a saddle

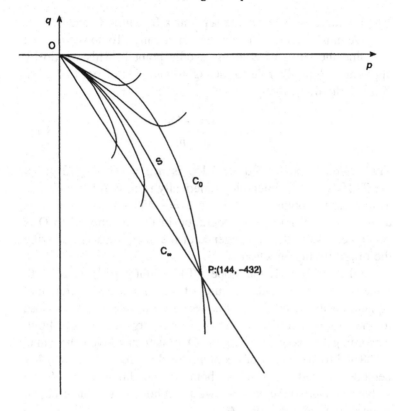

Figure 3.3.1. The fourth quadrant of the direction field of differential equation (3.3.14c). C_0 is the locus of zero slope and C_∞ is the locus of infinite slope.

point at P: $(144, -432)$. In the lenticular region between these two loci, the slope $dq/dp < 0$; to either side of this region $dq/dp > 0$ as shown in the diagram. A few integral curves are sketched in the figure.

Since $y(0)$ and $\dot{y}(0)$ are finite, when $x = 0$, $p = 0$ and $q = 0$. Thus the origin O lies on the integral curve we seek in the (p, q)-plane. But since O is a singular point, many integral curves pass through it. Now the asymptotic behavior $y = 144x^{-3}$ corresponds to the single point P in the (p, q)-plane. Thus the integral curve we want is the separatrix S joining O and P.

We must calculate this curve numerically. To do so we must first find the slope of S at the saddle point P. This we do by applying l'Hôpital's rule to the right-hand side of Eq. (3.3.14c). If $m = (dq/dp)_P$, then

$$m = \frac{4m + 18}{3 + m} \qquad (3.3.15)$$

The negative root of Eq. (3.3.15) is $m_- = (1 - \sqrt{73})/2 = -3.772\,002\ldots$. To integrate numerically from P to O we choose as our starting point $p = 144 - \varepsilon, q = -432 - m_-\varepsilon$, where ε is a suitably small number. Integration in the direction P to O is stable because of the convergence of the integral curves entering the origin in this direction.

When $y(0) = 1$, the slope $\dot{y}(0) = \lim(q/p^{4/3})$ as $p \to 0$. Because of the singular nature of the associated differential equation near the origin O, it is necessary to integrate very close to the origin in order to get good convergence of this limit. According to Davis [Da-62], E. D. Baker has given the value 1.588 588 to six places of decimals for this limit. Once $\dot{y}(0)$ has been determined, we have two boundary conditions at $x = 0$ and so have converted the original two-point boundary-value problem to an initial-value problem.∎

3.4 Singularities of the Associated Differential Equation

In the example of the Thomas–Fermi equation, Eq. (3.3.12), we have seen how the singularities of the associated differential equation played an important role. In this section, we prove two related theorems that apply when the singularities are saddle points. These theorems describe certain useful properties of the solutions $y(x)$ whose images in the (p, q)-plane are the separatrices passing through the saddle points.

The power-law solution $y = Ax^\beta$ to Eq. (3.3.4) (for which $\dot{y} = \beta Ax^{\beta-1}$) corresponds to the single point $(p, q) = (A, \beta A)$ if we take $p = y/x^\beta$ and $q = \dot{y}/x^{\beta-1}$. Now

$$x\frac{dq}{dx} = f(q, p) \tag{3.4.1a}$$

$$x\frac{dp}{dx} = q - \beta p \tag{3.4.1b}$$

where the function $f(q, p)$ is determined by the original differential equation (3.3.4) in x and y. For the power-law solution $y = Ax^\beta$, neither p nor q varies as x varies. Therefore the left-hand sides of Eqs. (3.4.1a–c) vanish. Now when $p = A$ and $q = \beta A$, the right-hand side of Eq. (3.4.1b) vanishes, so A must be determined by the equation $f(A, \beta A) = 0$. Since

$$\frac{dq}{dp} = \frac{f(q, p)}{q - \beta p} \tag{3.4.1c}$$

it is clear that the point P: $(A, \beta A)$ is a singular point in the (p, q)-plane.

If the singular point P is a saddle point, two separatrices pass through it. In the neighborhood of P, each separatrix can be approximated by a straight line

$$q - \beta A = m(p - A) \tag{3.4.2}$$

where m is the slope of the separatrix at P. Then on such a separatrix (which is an integral curve in its own right), Eq. (3.4.1b) can be written

$$\frac{dx}{x} = \frac{dp}{q - \beta p} = \frac{dp}{(m - \beta)(p - A)} \tag{3.4.3}$$

Equation (3.4.3) can be integrated to give

$$x = \text{const} \times (p - A)^{1/(m-\beta)} \tag{3.4.4}$$

Thus

Theorem 3.4.1.

> If $m > \beta$, then $x \to 0$ as $p \to A$, whereas if $m < \beta$, then $x \to \infty$ as $p \to A$.

We can see now from figure 3.3.1 (in which $\beta < 0$) that the separatrix S has a more negative slope m than the slope β of C_∞: $q - \beta p = 0$. Thus $m < \beta$ and as we approach P along the separatrix S, $x \to \infty$. Thus we determine S as the integral curve we want in the (p, q)-plane and furthermore determine that the solutions $y(x)$ that correspond to it have the asymptotic behavior $y = Ax^\beta$.

It may happen occasionally that it proves more convenient to take $p = F(y/x^\beta)$ and $q = G(y/x^\beta, \dot{y}/x^{\beta-1})$, where F and G are functions suitably chosen to make the associated differential equation take a convenient form. Then Eqs. (3.4.1a–c) take the more general form

$$x\frac{\mathrm{d}q}{\mathrm{d}x} = h(q, p) \qquad (3.4.5a)$$

$$x\frac{\mathrm{d}p}{\mathrm{d}x} = g(q, p) \qquad (3.4.5b)$$

so that

$$\frac{\mathrm{d}q}{\mathrm{d}p} = \frac{h(q, p)}{g(q, p)} \qquad (3.4.5c)$$

The singular point P that now corresponds to the power-law solution has the coordinates $(p_P, q_P) = (F(A), G(A, \beta A))$.

If the singular point is a saddle point, we again approximate its separatrices near P with the straight lines

$$q - q_P = m(p - p_P) \qquad (3.4.6)$$

Then since $g(q_P, p_P) = 0$, Eq. (3.4.5b) can be written near P as

$$\frac{\mathrm{d}x}{x} = \frac{\mathrm{d}p}{g_p(p - p_P) + g_q(q - q_P)} \qquad (3.4.7a)$$

which becomes on a separatrix

$$\frac{\mathrm{d}x}{x} = \frac{\mathrm{d}p}{(g_p + mg_q)(p - p_P)} \qquad (3.4.7b)$$

Here the partial derivatives g_p and g_q are evaluated at P. Integrating Eq. (3.4.7b) we find

$$x = \text{const} \times (p - p_P)^{1/(g_p + mg_q)} \qquad (3.4.8)$$

Now $g_p + mg_q$ is the directional derivative of the function $g(q, p)$ along the separatrix S. At the point P, $g(q, p) = 0$. Thus if $g_p + mg_q > 0$, then as we pass along S through P in the direction of increasing p, we pass from negative values of g to positive values of g. Furthermore, if $g_p + mg_q > 0$, then $x \to 0$ as $p \to p_P$. On the other hand, if $g_p + mg_q < 0$, then as we pass along S through P in the direction of increasing p, we pass from positive values of g to negative values of g. Furthermore, if $g_p + mg_q < 0$, then $x \to \infty$ as $p \to p_P$. To summarize:

Theorem 3.4.2a.

In the (p, q)-plane, as we pass along S through P in the direction of increasing p
    if g increases, $x = 0$ at P
    if g decreases, $x = \infty$ at P.

In the case $g(q, p) = q - \beta p$ as in Eq. (3.4.1b), this result is identical with the earlier result that if $m > \beta$, then $x \to 0$ as $p \to A$, whereas if $m < \beta$, then $x \to \infty$ as $p \to A$. Similarly,

Theorem 3.4.2b.

In the (p, q)-plane, as we pass along S through P in the direction of increasing q
    if h increases, $x = 0$ at P
    if h decreases, $x = \infty$ at P.

Using these rules, we can at once select the curve S joining O to P in figure 3.3.1 as the integral curve we want in the (p, q)-plane since O must correspond to $x = 0$ and P to $x = \infty$. The procedure outlined in this section is an alternative to the procedure of section 3.3.

3.5   A Caution in Applying Theorems 3.4.1 and 3.4.2

One must be careful in applying theorems 3.4.1 and 3.4.2 to be sure that the singularity arises from the simultaneous vanishing of the numerator and denominator of Eq. (3.4.1c) and is not a point at which the numerator becomes infinite. An example in which precisely this happens is the differential equation

$$3\frac{d(y\dot{y})}{dx} + 2x\dot{y} - y = 0 \qquad (3.5.1)$$

which arises in the theory of nonlinear diffusion. Again we seek solutions positive on the half-line $x \geq 0$ that vanish at infinity.

Equation (3.5.1) is invariant to the stretching group

$$x' = \lambda x \qquad (3.5.2a)$$
$$y' = \lambda^2 y \qquad (3.5.2b)$$

so that $\beta > 0$. Thus the solutions we seek cannot vanish at infinity proportionally to x^β. Furthermore, the power-law solution has $A = -1/6$ and thus cannot represent the asymptotic limit of a positive solution.

We can find the solutions we seek from an analysis of the direction field of the associated differential equation in the (p, q)-plane. If we set $p = y/x^2$ and $q = \dot{y}/x$, we find

$$x\frac{dq}{dx} = f(q, p) = \frac{p - 2q - 3q^2 - 3pq}{3p} \qquad (3.5.3a)$$

$$x\frac{dp}{dx} = q - 2p \qquad (3.5.3b)$$

so that

$$\frac{dq}{dp} = \frac{p - 2q - 3q^2 - 3pq}{3p(q - 2p)} \qquad (3.5.3c)$$

Figure 3.5.1 shows the fourth quadrant of the direction field of Eq. (3.5.3c). We are only interested in this quadrant because $p > 0$ and $q < 0$. (It follows from Eq. (3.5.1) that $\ddot{y} = 1/3$ at extrema, where $\dot{y} = 0$. Therefore, the only extrema possible

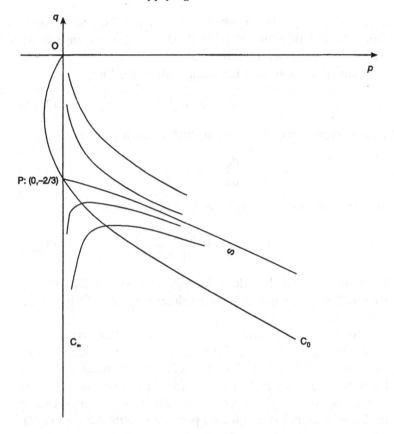

Figure 3.5.1. The fourth quadrant of the direction field of differential equation (3.5.3c). C_0 is the locus of zero slope and C_∞ is the locus of infinite slope.

are minima. Positive solutions that vanish at infinity cannot have only minima and hence must be monotone decreasing. Thus $\dot{y} \leq 0$.) The locus of zero slope C_0 is given by the equation

$$p = \frac{2q + 3q^2}{1 - 3q} \qquad (3.5.4)$$

The locus of infinite slope C_∞ has two branches, one in the first and third quadrants due to the vanishing of $q - 2p$ and a second, the q-axis, at which the function $f(q, p)$ becomes infinite. Thus the singular point P: $(0, -2/3)$ is *not* the kind of singularity discussed in section 3.4.

Now if we approach the point P along the line

$$q = mp - \tfrac{2}{3} \tag{3.5.5}$$

then it follows from Eq. (3.5.3a) that to leading order

$$x \frac{dq}{dx} = 1 + \frac{2m}{3} \tag{3.5.6}$$

as long as $m \neq -3/2$. Then, integrating, we have

$$x = x_0 \exp\left(\frac{q + 2/3}{1 + 2m/3}\right) \tag{3.5.7}$$

If we apply l'Hôpital's rule to the right-hand side of Eq. (3.5.3c), we find that the separatrix S has the slope $-3/4$ at P. Thus $x = x_0$, a constant, at P.

At $x = x_0$, $y(x_0)/x_0^2 = p_P = 0$ so that $y(x_0) = 0$. Furthermore, at $x = x_0$, $\dot{y}(x_0)/x_0 = q_P = -2/3$ so that $\dot{y}(x_0) = -2x_0/3$. These two conditions provide initial conditions for a *backward* integration of Eq. (3.5.1) starting from $x = x_0$ and going to $x = 0$. Because the slope $\dot{y}(x_0)$ is negative, a backward integration produces a positive solution on the interval $(0, x_0)$. Beyond x_0 the solution is negative and thus unacceptable. Hence, we have found solutions to Eq. (3.5.1) that are positive on the interval $(0, x_0)$ and fulfill the requirement that they vanish at infinity by vanishing for all $x \geq x_0$.

Since $y'(x_0') = 0$ and $\dot{y}'(x_0') = -2x_0'/3$, solutions that are distinguished from one another by different values of x_0 are images of one another under the transformations of the group (3.5.2). Hence, to find them all, we have to integrate Eq. (3.5.1) numerically only once. Now the point P in the (p, q)-plane corresponds to $x = x_0$, so integrating from $x = x_0$ towards $x = 0$ means moving along the separatrix S away from P in the direction of positive p. (Since $p = y/x^2$ and $q = \dot{y}/x$, if $y(0)$ and $\dot{y}(0)$ are finite, p and q become infinite as x approaches zero.)

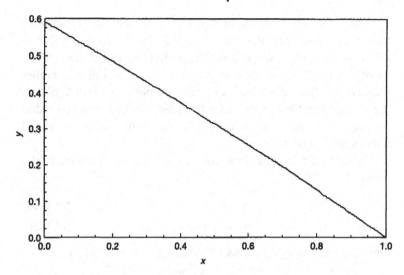

Figure 3.5.2. The integral curve of Eq. (3.5.1) for $x_0 = 1$ obtained by backward numerical integration using the boundary conditions $y(x_0) = 0$, $\dot{y}(x_0) = -2x_0/3$.

Because the integral curves converge in this direction, we expect the numerical integration to be stable. Shown in figure 3.5.2 is the integral curve for $x_0 = 1$ obtained in this way.

3.6 Other Groups

The examples of sections 3.2 to 3.5 all dealt with stretching groups, and sections 3.3 to 3.5 focused on problems on the semi-infinite interval $(0, \infty)$. The differential equation

$$x^{-\nu} \frac{\mathrm{d}}{\mathrm{d}x} \left(x^{\nu} \frac{\mathrm{d}y}{\mathrm{d}x} \right) = \mathrm{e}^{y} \qquad (3.6.1)$$

known as the Poisson–Boltzmann equation, provides an example of a problem on a finite interval in which the group is not a stretching group.

The Poisson–Boltzmann equation describes the concentration y of small, mobile ions (e.g. Na^+ ions) inside a solvent-filled cavity with walls charged oppositely to the charge of the mobile ions. The value of ν is 0, 1 or 2 according to whether the cavity is the slit between two charged planes perpendicular to the x-direction, the interior of a cylinder of which x is the radial coordinate, or the interior of a sphere of which x is the radial coordinate, respectively.

The Poisson–Boltzmann equation (3.6.1) is invariant to the group

$$x' = \lambda x \qquad\qquad (3.6.2a)$$

$$y' = y - 2\ln\lambda \qquad\qquad (3.6.2b)$$

the infinitesimal coefficients of which are $\xi = x$ and $\eta = -2$. As invariants we choose $p = x^2 e^y$ and $q = x\dot{y}$; then

$$\frac{dq}{dp} = \frac{p + (1 - \nu)q}{p(q + 2)} \qquad\qquad (3.6.3)$$

When $\nu = 1$ (cylindrical geometry), the variables p and q in Eq. (3.6.3) can be separated and Eq. (3.6.3) can be integrated at once to give

$$(q + 2)^2 = 2p + \text{constant} \qquad\qquad (3.6.4)$$

Now the solutions we seek are those for which the concentration of mobile ions at the origin is finite and has a finite derivative. Thus at $x = 0$, $p = q = 0$ so that the constant in Eq. (3.6.4) equals 4. Rewritten in terms of x, y and $\dot{y}$, Eq. (3.6.4) becomes

$$[(2x^2 e^y + 4)^{1/2} - 2]\,dx - x\,dy = 0 \qquad\qquad (3.6.5)$$

As noted in section 3.2, this first-order differential equation is also invariant to the group (3.6.2). Then according to Lie's theorem of section 2.1, an integrating factor is

$$(\xi M + \eta N)^{-1} = [x(2x^2 e^y + 4)^{1/2}]^{-1} \qquad\qquad (3.6.6)$$

Integration of Eq. (3.6.5) then yields

$$y = \ln\left[\frac{8b^2}{(b^2 - x^2)^2}\right] \qquad\qquad (3.6.7)$$

where b is an as yet undetermined constant of integration. The condition that fixes the constant b is the specification of the derivative dy/dx on the wall of the cavity, this derivative being determined by the surface charge density on the wall.

So far this example is not any more instructive than the example of the Emden–Fowler equation for $n = 5$ given in section 3.2. When $\nu = 0$ or $\nu = 2$, on the other hand, it appears that Eq. (3.6.3) cannot be integrated in terms of elementary functions. As most readers will recognize, when $\nu = 0$, Eq. (3.6.1) is a second-order differential equation in which the independent variable x does not appear explicitly. The classical procedure for such equations is to choose $q = \dot{y}$ and $p = y$ as new dependent and independent variables, respectively, in which case the differential equation reduces to one of first order in $\dot{y}$ and y. This is clearly an application of Lie's reduction theorem in which the underlying group is the translation group

$$y' = y \tag{3.6.8a}$$

$$x' = x + \lambda \tag{3.6.8b}$$

to which the differential equation (3.6.1) is also invariant (when $\nu = 0$). Thus when $\nu = 0$ the differential equation (3.6.1) is invariant to two groups.

The associated differential equation arising from the group (3.6.8) is easily solvable because, as we shall see in the next section, the group (3.6.2) supplies an integrating factor for it. A second integration is then possible, the integrating factor for it being supplied by the group (3.6.8). These integrations lead to the solution

$$y = \ln\left(\frac{c^2}{2}\sec^2\frac{cx}{2}\right) \tag{3.6.9}$$

that is nonsingular at the origin[1]. Here c is a constant of integration again related to the derivative dy/dx at the walls of the slit.

3.7   Equations Invariant to Two Groups

If a second-order differential equation is invariant to two groups, then its solution becomes a matter of quadrature, i.e. a matter of performing indefinite integrals. Let us see how this idea works out in the $v = 0$ case of Eq. (3.6.1). The invariants u of any group are the solutions of the partial differential equation (1.3.2), which we shall write here as $Uu = 0$, where U is the first-order differential operator

$$U = \xi \frac{\partial}{\partial x} + \eta \frac{\partial}{\partial y} \qquad (3.7.1)$$

The operator U is often called the *infinitesimal transformation* of the group because for any function $f(x, y)$

$$f(x', y') = f(x, y) + \left(\xi \frac{\partial f}{\partial x} + \eta \frac{\partial f}{\partial y} \right) \lambda \qquad (3.7.2a)$$

$$= f(x, y) + \lambda Uf \qquad (3.7.2b)$$

to first order in λ. Although U is called an infinitesimal transformation it does not contain the amplitude λ of the transformation as a factor. It is the directional derivative of f along the orbit of the group through the point (x, y):

$$\lim_{\lambda \to 0} \frac{f(x', y') - f(x, y)}{\lambda} = \xi \frac{\partial f}{\partial x} + \eta \frac{\partial f}{\partial y} = Uf \qquad (3.7.2c)$$

The infinitesimal transformation of U_1 of the (first) group (3.6.2) is thus

$$U_1 = x \frac{\partial}{\partial x} - 2 \frac{\partial}{\partial y} \qquad (3.7.3a)$$

whereas that of the (second) group (3.6.8) is

$$U_2 = \frac{\partial}{\partial x} \qquad (3.7.3b)$$

Now these infinitesimal operators can be multiplied by constants, added, and applied in succession. If we apply them in succession, then the order of the derivatives appearing increases, e.g.

$$U_1 U_2 = x \frac{\partial^2}{\partial x^2} - 2 \frac{\partial^2}{\partial y \partial x} \tag{3.7.4a}$$

$$U_2 U_1 = x \frac{\partial^2}{\partial x^2} + \frac{\partial}{\partial x} - 2 \frac{\partial^2}{\partial x \partial y} \tag{3.7.4b}$$

If we subtract Eq. (3.7.4a) from Eq. (3.7.4b), we eliminate the second derivatives and obtain

$$U_2 U_1 - U_1 U_2 = \frac{\partial}{\partial x} = U_2 \tag{3.7.5}$$

If p is an invariant of group 2, $U_2 p = 0$. Equation (3.7.5) shows that $U_2(U_1 p) = 0$ so that $U_1 p$ is an invariant of group 2. Therefore $U_1 p$ can be written as a function of p; call it $\Xi(p)$. Similarly, if q is a first differential invariant of group 2, $U_2 q = 0$, $U_2(U_1 q) = 0$, $U_1 q$ is a first differential invariant of group 2, and $U_1 q$ can be written as a function of p and q; call it $\mathrm{H}(p, q)$. Thus for any function $f(p, q)$ of p and q,

$$U_1 f = (U_1 p) \frac{\partial f}{\partial p} + (U_1 q) \frac{\partial f}{\partial q}$$

$$= \Xi(p) \frac{\partial f}{\partial p} + \mathrm{H}(p, q) \frac{\partial f}{\partial q} \tag{3.7.6}$$

Thus $\Xi(p)$ and $\mathrm{H}(p, q)$ are the infinitesimal coefficients of U_1 written in terms of the variables p and q.

The associated differential equation in the invariants p and q of group 2 can be written as $M(p, q) \, dp + N(p, q) \, dq = 0$. Since this associated differential equation is in reality another way of writing the original second-order differential equation, it must also be invariant to group 1. Therefore the quantity $(\Xi M + \mathrm{H} N)^{-1}$ must be an integrating factor for the associated differential equation $M(p, q) \, dp + N(p, q) \, dq = 0$ of group 2.

Let us now work these things out in detail for the infinitesimal operators in Eq. (3.7.3). Invariants of group 2 are

$p = y$ and $q = \dot{y}$ ($\xi = 1, \eta = 0, \eta_1 = 0$). Thus

$$\Xi(p) = U_1 p = U_1 y = \left(x \frac{\partial}{\partial x} - 2 \frac{\partial}{\partial y} \right) y = -2 \qquad (3.7.7a)$$

$$H(p, q) = U_1 q = U_1 \dot{y} = \left(x \frac{\partial}{\partial x} - 2 \frac{\partial}{\partial y} - \dot{y} \frac{\partial}{\partial \dot{y}} \right) \dot{y} = -\dot{y} = -q$$

$$(3.7.7b)$$

The associated differential equation in the variables $p = y$ and $q = \dot{y}$ is

$$e^p \, dp - q \, dq = 0 \qquad (3.7.8)$$

Now this differential equation is immediately integrable, but our goal here is to verify that the infinitesimal coefficients $\Xi(p)$ and $H(p, q)$ of U_1 supply an integrating factor, which should be $(-2e^p + q^2)^{-1}$ according to Lie's theorem. A short calculation shows that Eq. (3.7.8) can be integrated using this integrating factor to give $\phi(p, q) = $ constant, where

$$\phi(p, q) = -\tfrac{1}{2} \ln(q^2 - 2e^p) \qquad (3.7.9a)$$

Thus

$$q^2 - 2e^p = e^{-2\phi} = \text{constant} \qquad (3.7.9b)$$

which also follows directly from Eq. (3.7.8).

Equation (3.7.9b) is equivalent to

$$\dot{y}^2 - 2e^y = -c^2 \qquad (3.7.10)$$

where the constant has been written as $-c^2$. If the origin $x = 0$ is chosen to lie in the midplane between the walls of the slit, the solution we seek is symmetrical. Then, because $\dot{y}(0) = 0$ by symmetry and $e^y > 0$, $c^2 > 0$ and c is real. Since $p = y$ and $q = \dot{y}$ are invariants of group 2, Eq. (3.7.10), which is the same as Eq. (3.7.9b), is invariant to group 2. According to theorem 2.5.2, 'for groups in which the transformation of x is a translation, introducing a group invariant as a new variable and leaving the variable x unchanged also results in a new differential equation that is separable'. Equation (3.7.10) is separable just as expected from this statement, and integrating it yields Eq. (3.6.9).

3.8 Two-Parameter Groups

The key point in the argument of the last section is Eq. (3.7.5), which says that the quantity $U_2 U_1 - U_1 U_2$, called the *commutator* of U_1 and U_2, is equal to one of them. The argument would even apply if the commutator $U_2 U_1 - U_1 U_2$ were equal to a linear combination $\alpha U_1 + \beta U_2$ of U_1 and U_2. Here α and β are constants, either or both of which can be zero. If we set $V = \alpha U_1 + \beta U_2$, then

$$U_1 V - V U_1 = -\beta V \qquad (3.8.1a)$$

and

$$U_2 V - V U_2 = \alpha V \qquad (3.8.1b)$$

Furthermore, we can multiply Eq. (3.8.1a) by γ and Eq. (3.8.1b) by δ and add, obtaining

$$W V - V W = (\alpha\delta - \gamma\beta) V \qquad (3.8.1c)$$

where $W = \gamma U_1 + \delta U_2$. Using any of the equations (3.8.1) as a starting point, we can carry through the procedure of the previous section.

Note that not every pair of infinitesimal transformations has a commutator that is a linear combination with constant coefficients of the pair. For example, if

$$U_1 = x\frac{\partial}{\partial x} - by\frac{\partial}{\partial y} \qquad (3.8.2a)$$

and

$$U_2 = \frac{\partial}{\partial x} - c\frac{\partial}{\partial y} \qquad (3.8.2b)$$

then

$$U_2 U_1 - U_1 U_2 = \frac{\partial}{\partial x} + bc\frac{\partial}{\partial y} \qquad (3.8.2c)$$

which is only proportional to a linear combination of U_1 and U_2 (with constant coefficients) if $c = 0$ or $b = -1$.

Lie proved that the commutator of U_1 and U_2 is a linear combination with constant coefficients of U_1 and U_2 if and only if U_1 and U_2 are the infinitesimal transformations of a *two-parameter group*, $x' = X(x, y; \lambda, \mu)$, $y' = Y(x, y; \lambda, \mu)$. Indeed, he proved a generalization of this theorem for transformation groups that depend on more than two parameters and involve more than two variables. If we look upon the commutator as a generalized product, the set of infinitesimal transformations is then closed under addition, multiplication and multiplication by constants. Such a structure is known as an *algebra*, and the algebras of the infinitesimal transformations are called *Lie algebras*. Lie algebras have an extensive literature, but in this introductory book we go no further into their theory. For additional information regarding their application to differential equations, the interested reader may consult *Symmetries and Differential Equations* by G. W. Bluman and S. Kumei (Springer, New York, 1989). For applications in quantum theory, the reader may see *Classical Groups for Physicists* by B. G. Wybourne (Wiley–Interscience, New York, 1974) or *Linear Algebra and Group Theory* by V. I. Smirnov (McGraw-Hill, New York, 1961, translator: R. A. Silverman).

★3.9   Noether's Theorem[2]

If the first-order differential equation that arises from application of Lie's reduction theorem (the associated differential equation) can be integrated explicitly, the resulting solution, represented here by Eq. (3.2.4), is a function of x, y and $\dot{y}$ that is constant on the as yet unknown integral curves. Any such function of x, y and $\dot{y}$ that is constant on the integral curves is called a *first integral* of the differential equation. Because it is not always possible to integrate the associated differential equation explicitly, it is not always possible to display a first integral.

Noether has identified a class of differential equations for which it is always possible to display an explicit first integral.

These differential equations arise in the calculus of variations†
and are the Euler–Lagrange equations of invariant functionals.
We begin by considering the functional

$$J = \int_a^b L(x, y, \dot{y}) \, dx \tag{3.9.1}$$

which we suppose is invariant to the group (1.1.1) for all intervals
(a, b). (The quantities a and b, being values of x, transform
according to Eq. (1.1.1a).) Invariance of the functional means
that

$$J = \int_{a'}^{b'} L(x', y', \dot{y}') \, dx' \tag{3.9.2}$$

where J is the same number as in Eq. (3.9.1).

If we now change the variable of integration in the integral
to x, we obtain

$$J = \int_a^b L(x', y', \dot{y}') \frac{dx'}{dx} dx \tag{3.9.3}$$

where x', y' and $\dot{y}'$ are given, respectively, by Eqs. (1.1.1a),
(1.1.1b) and (1.5.4), and dx'/dx is given by Eq. (1.5.3) as

$$\frac{dx'}{dx} = X_x + \dot{y} X_y \tag{3.9.4}$$

Differentiating Eq. (3.9.3) with respect to λ and then setting
$\lambda = \lambda_0$, we find

$$0 = \int_a^b \left(\xi L_x + \eta L_y + \eta_1 L_{\dot{y}} + \frac{d\xi}{dx} L \right) dx \tag{3.9.5}$$

---

† The reader unfamiliar with the calculus of variations may consult Appendix C for a
brief introduction.

Since Eq. (3.9.5) is an identity in a and b, its integrand must vanish. Therefore

$$\xi L_x + \eta L_y + \left(\frac{d\eta}{dx} - \dot{y}\frac{d\xi}{dx}\right) L_{\dot{y}} + \frac{d\xi}{dx} L = 0 \qquad (3.9.6)$$

where we have used Eq. (1.5.5) for η_1. Equation (3.9.6) is the condition on the Lagrangian $L(x, y, \dot{y})$ for the functional (3.9.1) to be invariant.

The trajectories $y(x)$ that make the functional (3.9.1) an extremal obey the Euler–Lagrange differential equation

$$\frac{d(L_{\dot{y}})}{dx} - L_y = 0 \qquad (3.9.7)$$

On such a trajectory, Eq. (3.9.6) becomes

$$\boxed{\frac{d}{dx}[\eta L_{\dot{y}} + \xi(L - \dot{y}L_{\dot{y}})] = 0} \qquad (3.9.8)$$

so that the bracketed quantity $N(x, y, \dot{y})$ is a first integral of the Euler–Lagrange equation (3.9.7).

The characteristic equations of the partial differential equation (3.9.6) are

$$\frac{dx}{\xi} = \frac{dy}{\eta} = \frac{d\dot{y}}{\eta_1} = -\frac{dL}{L\,d\xi/dx} \qquad (3.9.9)$$

the integrals of which are $p(x, y)$, an invariant, $q(x, y, \dot{y})$, a first differential invariant, and ξL. Thus the most general form for the Lagrangian that makes the functional (3.9.1) invariant is

$$\boxed{L(x, y, \dot{y}) = \frac{G(p, q)}{\xi}} \qquad (3.9.10)$$

where G is a arbitrary function.

Example: The classical example of Noether's theorem is ordinary mechanics, for which the Lagrangian L is

$$L = \frac{m\dot{y}^2}{2} - U(y) \qquad (3.9.11)$$

The Euler–Lagrange differential equation of this Lagrangian is

$$m\ddot{y} + \dot{U}(y) = 0 \qquad (3.9.12)$$

A group that leaves the functional invariant is the translation group $x' = x + \lambda$, $y' = y$ for which $\xi = 1$, $\eta = 0$ and $\eta_1 = 0$. Thus $p = y$ and $q = \dot{y}$ are suitable invariants and L must have the form $G(y, \dot{y})$ as does Eq. (3.9.11). Noether's first integral is then

$$L - \dot{y}L_{\dot{y}} = -\frac{m\dot{y}^2}{2} - U(y) \qquad (3.9.13)$$

the constancy of which along a trajectory is simply the law of conservation of energy.■

Example: We can find the first integral (3.2.11) for the Emden–Fowler equation (3.2.5) when $n = 5$ using Noether's theorem. The Emden–Fowler equation (3.2.5) is the Euler–Lagrange equation of the Lagrangian

$$L = \frac{x^2\dot{y}^2}{2} - \frac{x^2 y^{n+1}}{n+1} \qquad (3.9.14)$$

Can we write this in the form (3.9.10) if the group is the stretching group (3.2.6)? Then L must have the form

$$L = \frac{G(y/x^\beta, \dot{y}/x^{\beta-1})}{x} \qquad (3.9.15)$$

For these last two equations to be the same, G must be chosen equal to

$$G\left(\frac{y}{x^\beta}, \frac{\dot{y}}{x^{\beta-1}}\right) = \frac{\dot{y}^2}{2x^{2(\beta-1)}} - \frac{y^{n+1}}{(n+1)x^{\beta(n+1)}} \qquad (3.9.16)$$

We see that $\beta(n+1)+1$ must equal -2 and $2\beta-1$ must equal -2 for the powers of x in each term of Eq. (3.9.14) to be the same as in the corresponding term in Eq. (3.9.15). Thus $\beta = -1/2$ and $n = 5$. Noether's first integral for this case is then

$$-\frac{x^2\dot{y}y}{2} - \frac{x^3\dot{y}^2}{2} - \frac{x^3 y^6}{6} \qquad (3.9.17)$$

which is the same result as given in Eq. (3.2.11).■

★3.10   Tabulation of Differential Equations Using Noether's Theorem

We can use Noether's theorem to tabulate differential equations for which an explicit first integral can be displayed. If Noether's first integral (3.9.8) is also a group invariant, it can be integrated again by the methods of chapter 2. Thus the solution of the tabulated differential equation can be reduced to a single integration.

In the preceding two examples, Noether's first integral N is a group invariant, but this is coincidental and not a general property of N. To see that N is a group invariant in the two examples, let us begin by substituting Eq. (3.9.10) for L into the bracketed expression for N in Eq. (3.9.8):

$$N = G + \left[\left(\frac{\eta}{\xi} - \dot{y} \right) q_{\dot{y}} \right] G_q \qquad (3.10.1)$$

Since G and G_q, being functions of the invariants p and q only, are, by their construction, invariants, N will be an invariant if the bracketed quantity in Eq. (3.10.1) is invariant.

In the first example, $\eta = 0$, $\xi = 1$ and $q = \dot{y}$. The bracketed quantity in Eq. (3.10.1) equals $-q$, which is clearly invariant. So in this case, $N = G - qG_q$, an invariant.

For the stretching group considered in the second example, $\xi = x$ and $\eta = \beta y$, $p = y/x^\beta$ and $q = \dot{y}/x^{\beta-1}$. Then the bracketed quantity in Eq. (3.10.1) equals $\beta p - q$, and $N = G + (\beta p - q)G_q$, an invariant.

On the other hand, for the rotation group (1.1.2), $\xi = -y$ and $\eta = x$, $p = x^2 + y^2$ and $q = (y - x\dot{y})/(x + y\dot{y})$. Then the bracketed quantity in Eq. (3.10.1) equals $x/y + q$, which is not invariant because x/y is not.

It can be shown (see problem 3.16) that if ξ depends only on x, then N is an invariant.

The tabulation procedure begins by picking a group (ξ, η). For the sake of a definite example let us choose the stretching group (3.2.6). Then the Euler–Lagrange equation (3.9.7) corresponding to the Lagrangian (3.9.15) is

$$\ddot{y} = x^{\beta-2}\left[\frac{G_p}{G_{qq}} + \frac{\beta G_q}{G_{qq}} + \frac{(\beta p - q)G_{pq}}{G_{qq}} + (\beta - 1)q\right] \quad (3.10.2)$$

where $p = y/x^{\beta}$ and $q = \dot{y}/x^{\beta-1}$. Equation (3.10.2) represents a class of differential equations of second order, invariant to the stretching group (3.2.6). For these differential equations, Noether's first integral is

$$N = G + (\beta p - q)G_q \quad (3.10.3)$$

and thus is an invariant. This means that Noether's first integral is an invariant first-order differential equation that can be treated by the methods of chapter 2. Thus the solution of the differential equation (3.10.2) can be reduced to a single quadrature (integration). As we have seen above, for the Emden–Fowler equation of order 5, $\beta = -1/2$ and $G = q^2/2 - p^6/6$.

Another group for which a similar conclusion holds is the mixed stretching–translation group

$$x' = \lambda x \quad (3.10.4a)$$
$$y' = y + \beta \ln \lambda \quad (3.10.4b)$$

For this group, $\xi = x, \eta = \beta, p = e^y/x^{\beta}$ and $q = x\dot{y}$. Then

$$N = G + (\beta - q)G_q \quad (3.10.5)$$

which is invariant. The Euler–Lagrange equation (3.9.7) is

$$G_{qq}x\ddot{y} + (G_{qq} + pG_{qp})\dot{y} - (\beta G_{qp} + G_p)\frac{p}{x} = 0 \quad (3.10.6)$$

The special case of Eq. (3.10.6) when $\beta = -2$ and $G = q^2/2 + p$ is the Poisson–Boltzmann equation (3.6.1) when $\nu = 1$.

3.11 The Determining Equations

One question that has been dealt with only obliquely so far is how to find groups to which a differential equation is invariant.

In section 2.6, Lie's method of tabulation was mentioned, and in principle this method could be used to create an extensive dictionary of differential equations that are invariant to various groups. Again in the previous section, 3.10, Noether's theorem was used to tabulate differential equations that were not only invariant to particular groups but for which a first integral could be explicitly displayed.

In the method of tabulation, one starts with a group and determines the invariant differential equations that belong to it. But in practice, the differential equation is forced upon us, so to speak, by other considerations, and if it does not happen to be tabulated already, how can we find groups to which it is invariant? There is a systematic approach to answering this question, but it involves extensive computation, especially for partial differential equations. In the remainder of this section, the basic ideas of the method are explained by working the example of the simple differential equation $\ddot{y} = 0$.

For this differential equation, Eq. (3.1.2a) becomes $\eta_2 = 0$. Now using Eq. (1.6.5), we find

$$\eta_2 = \eta_{xx} + \dot{y}(2\eta_{xy} - \xi_{xx}) + \dot{y}^2(\eta_{yy} - 2\xi_{xy}) - \dot{y}^3(\xi_{yy})$$
$$+ \ddot{y}(\eta_y - 2\xi_x - 3\dot{y}\xi_y) \qquad (3.11.1)$$

Setting $\ddot{y} = \eta_2 = 0$ gives

$$\eta_{xx} + \dot{y}(2\eta_{xy} - \xi_{xx}) + \dot{y}^2(\eta_{yy} - 2\xi_{xy}) - \dot{y}^3(\xi_{yy}) = 0 \quad (3.11.2)$$

Equation (3.11.2) is an identity in x, y and $\dot{y}$, that is, it holds for any arbitrary choice of x, y and $\dot{y}$. Since ξ and η are functions of x and y only, the coefficients of the various powers of $\dot{y}$ must vanish separately. Thus Eq. (3.11.2) is equivalent to the following four equations:

$$\eta_{xx} = 0; \qquad \xi_{yy} = 0; \qquad \xi_{xx} = 2\eta_{xy}; \qquad \eta_{yy} = 2\xi_{xy} \qquad (3.11.3)$$

These equations are called the *determining equations*.

According to the first two determining equations,

$$\xi = A(x)\, y + E(x) \qquad \text{and} \qquad \eta = I(y)\, x + O(y) \quad (3.11.4)$$

where A, E, I and O are functions yet to be determined. According to the second two determining equations

$$\ddot{A}(x)\, y + \ddot{E}(x) = 2\dot{I}(y) \qquad \text{and} \qquad \ddot{I}(y)\, x + \ddot{O}(y) = 2\dot{A}(x)$$
$$(3.11.5)$$

It follows by differentiating the first equation partially with respect to y and the second with respect to x that $\ddot{A}(x) = \ddot{I}(y) = 0$. Now because x and y are independent variables, it then follows that $\ddot{E}(x) = 2\dot{I}(y)$ and $\ddot{O}(y) = 2\dot{A}(x)$ are constants. Thus, finally

$$\xi = c_1 x^2 + c_2 xy + c_3 x + c_4 y + c_5 \qquad (3.11.6a)$$
$$\eta = c_2 y^2 + c_1 xy + c_6 x + c_7 y + c_8 \qquad (3.11.6b)$$

where the coefficients c_1 through c_8 are constants. The coefficient functions ξ and η belong to an eight-parameter group whose infinitesimal transformations (3.7.1) can be obtained by taking one of the coefficients c_1 through c_8 to be 1 and the rest zero.

Bluman and Kumei [Bl-89] have tabulated expressions for the coefficient functions of extended infinitesimal transformations of various orders for both one dependent and one independent variable and one dependent and several independent variables. These latter expressions are useful in obtaining the determining equations for partial differential equations.

Notes

Note 1: By working backwards from the solution (3.6.9), the author found the following solution in parametric form of Eq. (3.6.3) when $\nu = 0$, namely, $p = 2z^2 \sec^2 z$, $q = 2z \tan z$.

Note 2: The treatment of Noether's theorem given here follows that of Courant and Hilbert [Co-53, pp. 262–266] and is restricted, as is the rest of this introductory book, to point transformations in the (x, y)-plane. In her original paper [No-18], Noether considered more general transformations in which the infinitesimal coefficients ξ and η could depend not only on x and y but on $\dot{y}$ and higher derivatives as well. Furthermore, the transformations she considered were both multidimensional and multiparameter so that most generally the coefficients of the infinitesimal transformation depend on the independent variables, the dependent variables, and the partial derivatives of the dependent variables of both first and higher orders. Recently, even more general kinds of transformations (non-local transformations) have been studied in which the infinitesimal coefficients ξ and η depend on integrals of functions of x and y [Go-95].

It seems to me that in an introductory book such as this one, too high generality may complicate matters so much as to obscure the underlying idea. Therefore, the treatment in this book is restricted to the one-parameter, one-dimensional point transformations given in Eqs. (1.1.1a, b). It may be noted here, however, that the derivation in section 3.9 goes through without change even for more general one-parameter, one-dimensional transformations in which ξ and η depend not only on x and y but on $\dot{y}$ and higher derivatives as well. It follows again exactly as in section 1.6 that $\eta_1 = \mathrm{d}\eta/\mathrm{d}x - \dot{y}(\mathrm{d}\xi/\mathrm{d}x)$. With the help of Eq. (1.6.1a), we can pass at once from Eq. (3.9.3) to Eqs. (3.9.5) and (3.9.6) and from there to Eq. (3.9.8). Thus the bracketed quantity in Eq. (3.9.8) is again constant along a trajectory $y(x)$ that satisfies the Euler–Lagrange equation (3.9.7).

The interested reader may find quite complete treatments of Noether's theorem in her original paper [No-18] as well as in the books of G. W. Bluman and S. Kumei [Bl-89] and Peter J. Olver [Ol-86].

Problems for Chapter 3

3.1   The differential equation $1 + \dot{y}^2 = k^2(a-x)^2\ddot{y}^2$ arises in the calculation of curves of pursuit when the pursued travels along the straight line $x = a$ [Da-62]. Find a group to which it is invariant, use Lie's reduction theorem to reduce its order, and integrate twice to find the general solution.

3.2   Langmuir used the differential equation $3y\ddot{y} + \dot{y}^2 - e^x = 0$ in his theory of current flow from a hot cathode to an anode in a high vacuum [Da-62]. The differential equation has the singular solution $y = e^{x/2}$. Find this solution by finding a group to which Langmuir's differential equation is invariant, using Lie's reduction theorem, and determining the singular points of the associated equation.

3.3   What is the most general differential equation of the second order that is invariant to the stretching group $x' = \lambda x$, $y' = \lambda y$ and the translation group $x' = x + \lambda$, $y' = y + \lambda$?

3.4 Bessel's equation of order zero, $x^2\ddot{y} + x\dot{y} + x^2 y = 0$, like all homogeneous linear equations, is invariant to the stretching group $x' = x$, $y' = \lambda y$. Find an invariant u and a first differential invariant v and use Lie's reduction theorem to reduce the order of the differential equation. The associated equation is not, of course, integrable in terms of elementary functions, but when $x \gg 1$, it can be manipulated into an integrable form *correct to order* $1/x$. (*Hint*: derive a differential equation for $v + 1/(2x)$ by adding to or subtracting from the associated equation suitable terms of order $1/x^2$ or smaller.) Integrate the differential equation you have found, and then integrate again to find the asymptotic form of the Bessel functions of order zero.

3.5 Prove that the inhomogeneous linear differential equation of second order $\ddot{y} + P(x)\dot{y} + Q(x) y = R(x)$ is invariant to the group whose infinitesimal coefficients are $\xi = 0$, $\eta = Y(x)$, where $Y(x)$ is a particular solution of the homogeneous equation, i.e. where $\ddot{Y} + P(x)\dot{Y} + Q(x) Y = 0$. Find an invariant and a first differential invariant and use Lie's reduction theorem to reduce the order of the differential equation.

3.6 Find a stretching group to which the differential equation $x\ddot{y} - \dot{y} + 4x^2\dot{y}y = 0$ is invariant. Determine the form of possible power-law solutions and find them by direct substitution of this form into the differential equation. Use Lie's reduction theorem to reduce the differential equation to first order. Draw the direction field and using theorem 3.4.1 show that this power-law solution gives the asymptotic form of the family of solutions that obey the boundary conditions $y(0) > 0$, $y(\infty) = 0$.

3.7 Show that the solutions mentioned in problem 3.6 are ordered according to their values at $x = 0$.

3.8 The differential equation $d(y^2\dot{y})/dx + x\dot{y} = 0$, which arises in the applications of high-temperature superconductors, is invariant to the stretching group $x' = \lambda x$, $y' = \lambda y$. The quantity $u = y/x$ is an invariant and the quantity $v = \dot{y}$ is a first differential invariant. But any function of u is also an invariant and any function of u and v is also a first differential invariant. A more convenient choice is $u = y/x$ and $v = \dot{y}(y/x)^2$. Using these latter invariants, apply Lie's reduction theorem to obtain a first-order equation. In practice we are interested in positive, decreasing solutions $y(x)$ that vanish at infinity. In what part of the (u, v) direction field do such solutions lie? Consider the integral curve of the associated equation that passes through the origin and lies entirely in that part of the direction field. What can you say about the solutions $y(x)$ which correspond to it?

3.9 The differential equation $3\ddot{y} + x\dot{y}y - y^2 = 0$ arises in the study of the expulsion of a heated compressible fluid from a long, slender tube. We are interested in positive, decreasing solutions that vanish at infinity. Find a stretching group to which this differential equation is invariant, construct an invariant u and a first differential invariant v, and use Lie's reduction theorem to reduce the order of the equation. Draw the direction field of the associated equation and determine which integral curve in the (u, v)-plane corresponds to positive, decreasing solutions $y(x)$. What is the asymptotic form of these solutions? Prove that they are ordered according to their values at the origin. This being so, could you have found the asymptotic form without drawing the direction field?

3.10 Generalize the reasoning of section 3.2 to prove that if a third-order differential equation is invariant to a one-parameter group G, using an invariant p and a first differential invariant q as new variables reduces the differential equation to one of second order in p and q.

3.11 What is the most general differential equation of the second-order invariant to the group whose infinitesimal coefficients are $\xi = x^2$, $\eta = xy$? What variables can you use to reduce the order of the equation? What is the associated equation?

3.12 An old rule for reducing the order of second-order differential equations $\ddot{y} = f(x, y, \dot{y})$ is this: if either x or y is absent from f, choose the other as a new independent variable and choose $\dot{y}$ as a new dependent variable. Show that this rule follows from Lie's reduction theorem by choosing appropriate groups that leave the differential equation invariant.

3.13 Find two groups G_1 and G_2 to which the differential equation $x^2\ddot{y} + x\dot{y} - 1 = 0$ is invariant. Calculate their infinitesimal transformations U_1 and U_2 and show that they commute, i.e. that their commutator $U_1U_2 - U_2U_1 = 0$. Using G_1, reduce the order of the differential equation by means of Lie's reduction theorem. Then use G_2 to find Lie's integrating factor for the associated equation. Integrate the associated equation and find a first integral of the original second-order differential equation. Show that this first integral is invariant to G_1. Integrate it to find the general solution of the original second-order differential equation.

3.14 If we set $G = q^2/2 + \ln p$ in Eq. (3.10.6) we obtain the second-order differential equation of problem 3.13, namely, $x^2\ddot{y} + x\dot{y} - 1 = 0$. Calculate Noether's first integral (3.10.5). Is the value of β fixed in this particular problem? If not, what use can you make of its arbitrariness?

3.15 (Do this problem before problem 3.16!) Show that the quantity $(\eta/\xi - \dot{y})q_{\dot{y}}$ appearing in Eq. (3.10.1) is an invariant of any group for which ξ depends only on x and η depends only on y.

3.16 Prove that a necessary and sufficient condition for the quantity $(\eta/\xi - \dot{y})q_{\dot{y}}$ appearing in Eq. (3.10.1) to be a first differential invariant is that $\xi_y = 0$, i.e. that ξ depend only on x.

3.17

(a) Find the condition on ξ and η for the group of transformations (1.1.1a, b) to preserve area. (Preserving area means that if C is any closed curve in the (x, y)-plane and C' is its image for any fixed value of λ, then the area inside C' equals the area inside C.)

(b) Find the most general expressions for ξ and η of the area-preserving groups whose orbits are the family $u(x, y) = c$.

4

Similarity Solutions of Partial Differential Equations

4.1 Introduction

The aim in the foregoing chapters was to find the general solution of ordinary differential equations. Finding the general solution of most of the partial differential equations of science and engineering is too difficult. Instead, we are usually satisfied to find particular solutions defined by specific initial and boundary conditions. If the partial differential equation is invariant to a group, then among its solutions may be some that are their own images under transformation (in general, the image of a solution is another solution). These invariant solutions are generally easier to calculate than other solutions; often they may be calculated by solving an ordinary differential equation. If these invariant solutions also describe situations of practical interest, they may provide us with simple answers to seemingly difficult questions.

In this chapter, we confine our attention to certain partial differential equations in one dependent variable (call it c) and two independent variables (call them z and t) that are invariant to the following one-parameter family of one-parameter stretching groups:

$$c' = \lambda^\alpha c \qquad\qquad\qquad (4.1.1a)$$

$$t' = \lambda^\beta t \qquad 0 < \lambda < \infty \qquad (4.1.1b)$$

$$z' = \lambda z \qquad\qquad\qquad (4.1.1c)$$

As before, λ is the *group parameter* that labels different transformations of a group; α and β are *family parameters* that label different groups of the family. The partial differential equations considered here are invariant to every group in the family, but not every pair of values of α and β is admissible: for each partial differential equation, the family parameters α and β are coupled by the linear constraint

$$\boxed{M\alpha + N\beta = L} \qquad (4.1.2)$$

where M, N and L are fixed constants determined by the structure of the particular partial differential equation. Thus only one of the family parameters α and β can be chosen independently. As we shall see, the values of α and β in a particular problem are determined by the boundary and initial conditions.

The restrictions described above permit some useful theorems to be proved while still allowing many partial differential equations of practical interest to be treated. Among such partial differential equations are

(1) the ordinary diffusion equation $c_t = c_{zz}$,

(2) the heat diffusion equation in superfluid helium, $c_t = (c_z^{1/3})_z$;

(3) the equation $cc_t = c_{zz}$, which arises in the theory of expulsion of a compressible fluid from a long, heated tube;

(4) the equation $c_{tt} = p + c_{zz} \int_0^1 (c_z^2/2)\,\mathrm{d}z$, which arises in the theory of motion of a shock-loaded elastic membrane;

(5) the equation $c_t = (c^n)_{zz}$, which describes magnetic diffusion in high-temperature superconductors;

(6) the coupled partial differential equations $u_t = v_z$, $v_t = V^2 u_z$, which describe the stretching of a long, elastic wire; and

(7) the coupled equations $h_t + (vh)_z = 0$, $v_t + vv_z + h_z = 0$ of problem (4.1.2), which describe the motion of water in a long, narrow, open channel.

4.2   Similarity Solutions

Each solution $c(z, t)$ of a partial differential equation represents a surface S in (c, z, t)-space. If the partial differential equation is invariant under a group G, then the image surface of S, composed of the images (c', z', t') of the points (c, z, t) of S, generally represents another solution of the partial differential equation. But if the solution $c(z, t)$ itself is invariant under the transformations of G, then the surface S is its own image.

Let the surface S be represented by the equation

$$F(c, z, t) = 0 \qquad (4.2.1a)$$

If it is its own image, then

$$F(c', z', t') = 0 \qquad (4.2.1b)$$

as well. Equation (4.2.1b) can be written

$$F(\lambda^\alpha c, \lambda z, \lambda^\beta t) = 0 \qquad (4.2.2)$$

If we differentiate Eq. (4.2.2) with respect to λ and then set $\lambda = 1$, we obtain the linear partial differential equation

$$\alpha c F_c + z F_z + \beta t F_t = 0 \qquad (4.2.3)$$

whose characteristic equations are

$$\frac{dc}{\alpha c} = \frac{dz}{z} = \frac{dt}{\beta t} = \frac{dF}{0} \qquad (4.2.4)$$

Three independent integrals of Eqs. (4.2.4) are

$$F, x = z/t^{1/\beta} \quad \text{and} \quad y = c/t^{\alpha/\beta} \qquad (4.2.5)$$

The most general solution of Eq. (4.2.3) is $F = f(x, y)$, where f is any arbitrary function. Since the surface S is designated by $f(x, y) = F = 0$, we can solve this last equation for y in terms of x and write: $y = y(x)$ where $y(x)$ is an arbitrary function of x only. Written in terms of c, z, t, the relation $y = y(x)$ becomes

$$\boxed{c = t^{\alpha/\beta} y(z/t^{1/\beta})} \qquad (4.2.6)$$

which is the most general form of a solution which itself is invariant to a group of the family (4.1.1). A solution of the original partial differential equation of the form (4.2.6) is called a *similarity solution*.

If Eq. (4.2.6) is substituted into the partial differential equation, an ordinary differential equation for y must result because y is a function of only one argument, namely, $z/t^{1/\beta}$. A convenient name for this ordinary differential equation is *principal differential equation*. Because similarity solutions can be found by solving an ordinary differential equation, they are much simpler to calculate than other solutions of the partial differential equation. Unfortunately, not every problem, i.e. not every set of boundary and initial conditions, leads to a similarity solution. For example, one can see at once that the form (4.2.6) restricts the boundary value $c(0, t)$ to powers of t. Furthermore, if Eq. (4.2.6) is to represent a solution that spreads out as time increases, β must be positive. In spite of these restrictions, many problems of practical interest do lead to similarity solutions.

Example: The superfluid diffusion equation $c_t = (c_z^{1/3})_z$ is invariant to the family (4.1.1) with $M = 2$, $N = -3$ and $L = -4$, i.e. with $2\alpha - 3\beta = -4$. If we substitute Eq. (4.2.6) into this partial differential equation, we obtain the following ordinary differential equation for y:

$$\beta\frac{d(\dot{y}^{1/3})}{dx} + x\dot{y} - \alpha y = 0 \qquad (4.2.7)$$

where, as usual, $\dot{y}$ stands for the derivative of y with respect to its argument, that is, dy/dx. We investigate three different sets of boundary and initial conditions.

(a) The boundary and initial conditions $c(0, t) = 1$, $c(\infty, t) = 0$ and $c(z, 0) = 0$ correspond to what I call the clamped-temperature problem: here c is the temperature rise in a semi-infinite pipe initially filled with cold superfluid helium [$c(z, 0) = 0$] the temperature of whose front face is clamped so that $c(0, t) = 1$ for $t > 0$. The condition $c(0, t) = 1$

requires α to be zero, so that $\beta = 4/3$. Then the differential equation (4.2.7) takes the form (3.3.9) discussed previously.

(b) The boundary and initial conditions $c_z(0, t) = -1$, $c(\infty, t) = 0$ and $c(z, 0) = 0$ correspond to what I call the clamped-flux problem: here c is the temperature rise in a semi-infinite pipe initially filled with cold superfluid helium $[c(z, 0) = 0]$ the heat flux of whose front face is clamped so that $c_z(0, t) = -1$ for $t > 0$. The condition $c_z(0, t) = -1$ requires α to be 1, so that $\beta = 2$. Then the differential equation (4.2.7) takes the form

$$2\frac{d(\dot{y}^{1/3})}{dx} + x\dot{y} - y = 0 \qquad (4.2.8)$$

(c) The boundary and initial conditions $c(\infty, t) = 0$ and $c(z, 0) = 0$ and the conservation condition $\int_{-\infty}^{\infty} c\,dz = 1$ correspond to what I call the pulsed-source problem: here c is the temperature rise in an infinite pipe initially filled with cold superfluid helium $[c(z, 0) = 0]$ subjected to a unit heat pulse in the plane $z = 0$ at $t = 0$. The condition $\int_{-\infty}^{\infty} c\,dz = 1$ requires α to be -1, so that $\beta = 2/3$. Then the differential equation (4.2.7) takes the form

$$2\frac{d(\dot{y}^{1/3})}{dx} + 3(x\dot{y} + y) = 0 \qquad (4.2.9)$$

These three examples illustrate how the boundary and initial conditions determine the values of the family parameters α and β. Furthermore, the specific form of the principal differential equation changes from case to case: Eq. (4.2.7) is analytically integrable when $\alpha = 0$ or -1 but not when $\alpha = 1$. Finally, one should note that the principal differential equation (4.2.7) is invariant to the stretching group

$$y' = \lambda^{-2} y \qquad (4.2.10a)$$

$$x' = \lambda x \qquad (4.2.10b)$$

Thus it can be treated by the methods of chapter 3. This invariance is no coincidence as we see next.■

4.3 The Associated Group

The principal differential equation that arises from the family of stretching groups (4.1.1) is invariant to the *associated* stretching group

$$y' = \lambda^{L/M} y \qquad (4.3.1a)$$
$$x' = \lambda x \qquad (4.3.1b)$$

where L and M are two of the coefficients in the linear constraint (4.1.2). To prove this we follow the form of argument used in section 3.8.

The infinitesimal transformation X of the group G of the family (4.1.1) labeled by the family parameters α and β is

$$X = \alpha c \frac{\partial}{\partial c} + \beta t \frac{\partial}{\partial t} + z \frac{\partial}{\partial z} \qquad (4.3.2a)$$

The infinitesimal transformation X_*, defined by

$$X_* = \alpha_* c \frac{\partial}{\partial c} + \beta_* t \frac{\partial}{\partial t} + z \frac{\partial}{\partial z} \qquad (4.3.2b)$$

where $\alpha_* \neq \alpha$ and $\beta_* \neq \beta$ and $M\alpha_* + N\beta_* = L$, also leaves the differential equation invariant. A simple calculation shows that the commutator of X and X_* vanishes. Since $Xx = 0$ and $Xy = 0$ by the definition of x and y as invariants of X, it follows that $X(X_*x) = 0$ and $X(X_*y) = 0$. Therefore, X_*x and X_*y are invariants of X and so can be expressed as functions of x and y only. According to the chain rule, then, the action of X_* on any function of x and y is given by

$$X_* = (X_*x) \frac{\partial}{\partial x} + (X_*y) \frac{\partial}{\partial y} \qquad (4.3.3)$$

We can calculate the functions X_*x and X_*y explicitly from Eqs. (4.2.5) and (4.3.2b):

$$X_*x = \left(\frac{\beta - \beta_*}{\beta} \right) x \qquad (4.3.4a)$$

$$X_*y = \left(\frac{\alpha_* \beta - \alpha \beta_*}{\beta} \right) y = \left(\frac{\beta - \beta_*}{\beta} \right) \frac{L}{M} y \qquad (4.3.4b)$$

Thus

$$X_* = \left(\frac{\beta - \beta_*}{\beta} \right) \left(x \frac{\partial}{\partial x} + \frac{L}{M} y \frac{\partial}{\partial y} \right) \qquad (4.3.4c)$$

The transformations of the group G_* generated by X_* in (x, y)-space are obtained by solving the differential equations

$$\frac{dx}{x} = \frac{dy}{Ly/M} = \frac{d\lambda}{\lambda} \qquad (4.3.5)$$

They are Eqs. (4.3.1).

Because X_* is an infinitesimal transformation of the group (4.1.1), it must carry one solution of the partial differential equation into another. Therefore, the transformations of G_* in (x, y)-space carry one similarity solution $y(x)$ belonging to the family parameters α and β into another similarity solution $y'(x')$ belonging to the same family parameters. The totality of similarity solutions belonging to a particular set of family parameters then forms an invariant family. This demonstrates that the principal differential equation is invariant to the associated group (4.3.1) for any choice of the family parameters α and β.

4.4   The Asymptotic Behavior of Similarity Solutions

If $L/M < 0$ and if the positive solutions of the principal differential equation that vanish at infinity are ordered according to their values at $x = 0$, then by the theorem of section 3.3 the positive similarity solutions $y(x)$ that vanish at infinity are asymptotic to the exceptional solution $y = Ax^{L/M}$ with the smallest positive value of A. In terms of c, z and t, this asymptotic behavior can be written $c = Az^{L/M} t^{-N/M}$. The value of A can be found directly by substituting this last form into the partial differential equation.

The solutions $y(x)$ of the principal differential equations belonging to the illustrative partial differential equations (1), (2) and (3) given at the end of section 4.1, are ordered according to their values at $x = 0$. This is because the solutions $c(z, t)$

of the partial differential equations obey the following *ordering theorem*.

Theorem 4.4.1.

> If $c_1(0, t) \geq c_2(0, t)$ and $c_1(\infty, t) \geq c_2(\infty, t)$ for $0 < t < T$, and $c_1(z, 0) \geq c_2(z, 0)$ for $0 < z < \infty$, then $c_1(z, t) \geq c_2(z, t)$ for $0 < t < T$ and $0 < z < \infty$.

We prove this theorem in the next section; here we take it for granted and explore its consequences.

The three problems discussed in the example at the end of section 4.2 all have the partial boundary and initial conditions $c(\infty, t) = 0$ and $c(z, 0) = 0$. For two solutions $c_1(z, t)$ and $c_2(z, t)$ fulfilling these conditions, if $c_1(0, t) \geq c_2(0, t)$ for $0 < t < T$, then $c_1(z, t) \geq c_2(z, t)$ for $0 < t < T$ and $0 < z < \infty$. If c_1 and c_2 are similarity solutions belonging to the same α and β, then $c_1(0, t) \geq c_2(0, t)$ means $y_1(0) \geq y_2(0)$ and $c_1(z, t) \geq c_2(z, t)$ means $y_1(x) \geq y_2(x)$. Thus the similarity solutions have the asymptotic behavior given in the first paragraph of this section, namely, $y = Ax^{L/M}$, i.e. $c = Az^{L/M}t^{-N/M}$.

The asymptotic behavior $c = Az^{L/M}t^{-N/M}$ extends to other solutions of the partial differential equation besides similarity solutions. By way of illustration, let us return to the superfluid diffusion equation discussed at the end of section 4.2. Since $M = 2$, $N = -3$, $L = -4$ and $A = 4/(3\sqrt{3})$, the similarity solutions defined by the partial boundary and initial conditions $c(\infty, t) = 0$ and $c(z, 0) = 0$ are asymptotic to $[4/(3\sqrt{3})]z^{-2}t^{3/2}$. Consider now the solution $c(z, t)$ defined by the boundary and initial conditions $c(\infty, t) = 0$, $c(z, 0) = 0$ and $c(0, t) = 1 + \tanh t$. Since $1 \leq 1 + \tanh t \leq 2$, by the ordering theorem $c(z, t)$ is sandwiched between the two clamped-temperature similarity solutions ($\alpha = 0$, $\beta = 4/3$) for which $c(0, t) = 1$ and $c(0, t) = 2$, respectively. Since both of these similarity solutions are asymptotic to $[4/(3\sqrt{3})]z^{-2}t^{3/2}$ so is $c(z, t)$.

There are some limitations on α and β imposed by the constraint (4.1.2) and by the requirement that β be > 0 in order that the similarity solution spread out as time increases (the usual case of interest). In the case of the superfluid diffusion equation, these constraints require that $\alpha > -2$. When $\beta > 0$ and $-2 < \alpha < \infty$, $-\infty < \alpha/\beta < 3/2$. Thus any solution for which $c(\infty, t) = c(z, 0) = 0$ and for which $c(0, t)$ can be sandwiched between two powers of t with exponents $< 3/2$ has the asymptotic behavior $[4/(3\sqrt{3})]z^{-2}t^{3/2}$.

★4.5 Proof of the Ordering Theorem

The illustrative partial differential equations (1), (2) and (3) given at the end of section 4.1 all belong to a class of equations called *conservation equations*, which have the form

$$\boxed{S(c)\, c_t + q_z = 0} \tag{4.5.1}$$

where, most generally, q is a function of z, t, c and c_z. To prove the ordering theorem of the last section, we further assume that $S(c) > 0$ and $\partial q/\partial c_z \leq 0$ and use the style of argument employed earlier in connection with Eq. (3.3.9). Let c_1 and c_2 be two infinitesimally close neighboring solutions of Eq. (4.5.1) and suppose that $c_1(0, t) \geq c_2(0, t)$ and $c_1(Z, t) \geq c_2(Z, t)$ for $0 < t < T$ and $c_1(z, 0) \geq c_2(z, 0)$ for $0 < z < Z$. Define $u = c_1 - c_2$. Then $u(0, t) \geq 0$ and $u(Z, t) \geq 0$ for $0 < t < T$ and $u(z, 0) \geq 0$ for $0 < z < Z$.

The difference u obeys the partial differential equation

$$S u_t = -\left(\frac{\partial q}{\partial c_z}\right) u_{zz} - \left[\frac{\partial}{\partial z}\frac{\partial q}{\partial c_z} + \frac{\partial q}{\partial c}\right] u_z - \left[\frac{dS}{dc}c_t + \frac{\partial}{\partial z}\frac{\partial q}{\partial c}\right] u \tag{4.5.2}$$

Equation (4.5.2) has been obtained by subtracting Eq. (4.5.1) written for c_2 from Eq. (4.5.1) written for c_1. The mean value theorem[1] has been used to evaluate the differences $S(c_1)\, c_{1t} - S(c_2)\, c_{2t}$ and $q_z(z, t, c_1, c_{1z}) - q_z(z, t, c_2, c_{2z})$. As a consequence, S, q and their derivatives are evaluated for arguments lying between c_1 and c_2. But because c_1 and c_2 are infinitesimally close, this causes no difficulty.

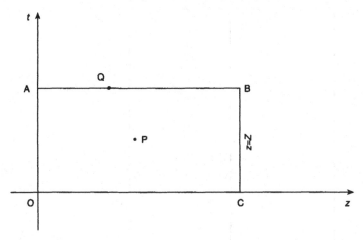

Figure 4.5.1. The rectangle used in the proof of the ordering theorem of section 4.5.

To show the ordering of c_1 and c_2 we must show that $u \geq 0$ everywhere in the rectangle OABC of figure 4.5.1. The proof depends on finding where the smallest value of u can lie. To clarify the logic of the argument let us first consider a function w of one variable s that is positive at the endpoints of an interval $a < s < b$ (see figure 4.5.2). If the smallest value of w on the interval occurs at an interior point P of the interval, then $(\mathrm{d}w/\mathrm{d}s)_{\mathrm{P}} = 0$ (curves 2 and 3). On the other hand, if $\mathrm{d}w/\mathrm{d}s$ does not vanish anywhere on the interval, then the smallest value of w occurs at an endpoint (curve 1). In case 2, where the smallest value is positive, $w > 0$ on the entire interval. In case 1, where the endpoint values are positive, w is likewise > 0 on the entire interval. What we try to do is use the partial differential equation to eliminate a case like 3, where the smallest value is negative. With these considerations in mind, we begin by trying to prove that u cannot attain its smallest value in the interior of rectangle OABC if that value is negative.

For if it were so u would have a negative minimum at a point P in the interior of rectangle OABC. Then $u_z(\mathrm{P}) = u_t(\mathrm{P}) = 0$, $u(\mathrm{P}) < 0$ and $u_{zz}(\mathrm{P}) \geq 0$. If the bracketed quantity in the

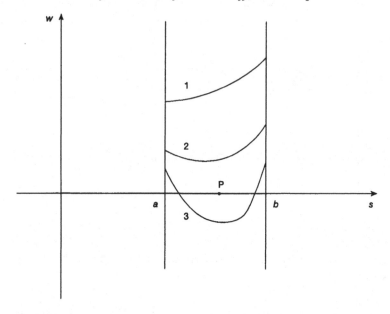

Figure 4.5.2. An auxiliary diagram used in the proof of the ordering theorem of section 4.5.

last term on the right-hand side of Eq. (4.5.2) were positive, these conditions would be inconsistent with the partial differential equation (4.5.2). This would provide a contradiction that refutes the assumption that u had a negative minimum at P.

Unfortunately, we do not know the sign of the bracketed quantity. But Protter and Weinberger [Pr-67] describe an artifice due to Hopf by which the foregoing argument can be saved. Let

$$v = ue^{-\gamma t} \qquad (4.5.3)$$

where γ is a positive constant yet to be determined. Then $v(0, t) \geq 0$ and $v(Z, t) \geq 0$ for $0 < t < T$ and $v(z, 0) \geq 0$ for $0 < z < Z$. Furthermore, v obeys the partial differential equation

$$Sv_t = -\left(\frac{\partial q}{\partial c_z}\right)v_{zz} - \left(\frac{\partial}{\partial z}\frac{\partial q}{\partial c_z} + \frac{\partial q}{\partial c}\right)v_z - \left(\frac{dS}{dc}c_t + \frac{\partial}{\partial z}\frac{\partial q}{\partial c} + \gamma S\right)v$$

$$(4.5.4)$$

Now since $S > 0$, if we choose γ large enough the bracketed quantity in the last term on the right is positive. Therefore, v cannot attain its smallest value in the interior of OABC if that value is negative.

Furthermore, v cannot attain its smallest value at an interior point Q of side AB if that value is negative. For then v would have a negative minimum at Q and thus $v_z(Q) = 0$, $v(Q) < 0$ and $v_{zz}(Q) \geq 0$. Then Eq. (4.5.4) requires that $v_t > 0$, which means that yet smaller values of v lie inside OABC just below Q, again a contradiction.

Now only two alternatives are left: either v attains its smallest value in the interior of OABC or in the interior of segment AB, in which case that smallest value is ≥ 0, or v attains its smallest value somewhere on sides OA, OC or CB, in which case that smallest value is again ≥ 0. Thus $v \geq 0$ everywhere in rectangle OABC and on its perimeter. Since $e^{-\gamma t} > 0$, $u \geq 0$ everywhere in rectangle OABC and on its perimeter as well, which was to be proved.

4.6 Functions Invariant to an Entire Family of Stretching Groups

Whereas the similarity solution (4.2.6) is invariant to one group of the family (4.1.1), the exceptional solution $c = Az^{L/M}t^{-N/M}$ is invariant to every group of the family. In other words, if we write $c' = Az'^{L/M}t'^{-N/M}$ and substitute Eqs. (4.1.1) for the primed variables, we obtain $c = Az^{L/M}t^{-N/M}$ for any α and β obeying Eq. (4.1.2).

Furthermore, the solution $c = Az^{L/M}t^{-N/M}$ is the only solution invariant to the entire family (4.1.1). To see the uniqueness of the solution $c = Az^{L/M}t^{-N/M}$, we need only assemble some results already obtained. In the first place, a totally invariant solution must be an invariant of both the infinitesimal transformation X of Eq. (4.3.2a) and the infinitesimal transformation X_* of Eq. (4.3.2b). Now the most general invariant of X is $F(x, y)$ as noted after Eq. (4.2.5). Since

$F(x, y)$ must be invariant to X_* given in Eq. (4.3.4c), it must have the form $G(y/x^{L/M})$ [cf. Eq. (3.2.7a)]. A totally invariant solution must therefore have the form $G(y/x^{L/M}) = 0$, which is equivalent to $y = Ax^{L/M}$ or $c = Az^{L/M}t^{-N/M}$.

Besides determining the asymptotic behavior of solutions under conditions already discussed, totally invariant functions can be used to determine other useful information. The following example occurs in the theory of stability of uncooled superconductors. Consider the one-dimensional heat diffusion equation with a temperature-dependent source on the infinite interval $-\infty < z < \infty$:

$$c_t = c_{zz} + Q(c) \tag{4.6.1}$$

Let the temperature rise $c(z, t) = 0$ for $t < 0$, and at $t = 0$ let a heat pulse q be introduced at $z = 0$. This initial condition of a localized pulsed source can be represented analytically by the requirement that for very short times, at which the source Q has yet to make itself felt,

$$c = q(4\pi t)^{-1/2}\exp\left(-\frac{z^2}{4t}\right) \tag{4.6.2}$$

For commercial superconductors, $Q(c)$ is often taken to be a step function

$$Q(c) = 0 \qquad c < a \tag{4.6.3a}$$
$$Q(c) = b \qquad c > a \tag{4.6.3b}$$

Now, if q is small enough, the temperature c eventually returns to zero, whereas if q is too large, c eventually diverges without limit. We want the *bifurcation* value of q, i.e. the largest value of q for which c still returns to zero.

The problem just posed is one of a class of problems, which class is invariant to the family of stretching groups

$$c' = \lambda^\delta c \qquad t' = \lambda^2 t \qquad z' = \lambda z \tag{4.6.4a}$$
$$q' = \lambda^{\delta+1}q \qquad b' = \lambda^{\delta-2}b \qquad a' = \lambda^\delta a \tag{4.6.4b}$$

For every problem of this class q is determined only by a and b. Therefore, $F(q, a, b) = 0$ for some function $F(q, a, b)$ that is the same for all problems of the class and thus is an invariant of the family of stretching groups given by Eq. (4.6.4b). If we set $\mu = \lambda^\delta$, we can write Eq. (4.6.4b) as

$$q' = \mu^\alpha q \qquad b' = \mu^\beta b \qquad a' = \mu a \qquad (4.6.4c)$$

where $\alpha = (\delta + 1)/\delta$ and $\beta = (\delta - 2)/\delta$ so that

$$2\alpha + \beta = 3 \qquad (4.6.4d)$$

Thus $M = 2$, $N = 1$ and $L = 3$. Therefore,

$$q = Aa^{L/M}b^{-N/M} = Aa^{3/2}b^{-1/2} \qquad (4.6.5)$$

where A is a constant independent of a and b that can be determined by solving the problem numerically for one particular choice of a and b. Thus we see from group-theoretic arguments that only one such solution is necessary to determine the dependence of q on a and b, whereas it appeared at the outset that the problem might have to be solved repeatedly for each choice of a and b.

★4.7 The Shock-Loaded Membrane

As noted at the end of section 4.1, the equation

$$c_{tt} = p + c_{zz} \int_0^1 \frac{c_z^2}{2} \, dz \qquad (4.7.1)$$

arises in the theory of motion of a shock-loaded elastic membrane. Shown in figure 4.7.1 is a cross-section of such a membrane in the form of a long elastic ribbon rigidly clamped at its sides $z = 0$ and $z = 1$. (We work here in special, rescaled units in which the mass per unit length of the membrane, its width, and its Young's modulus are all equal to 1. The derivation of Eq. (4.7.1) can be found in the author's earlier book [Dr-83].)

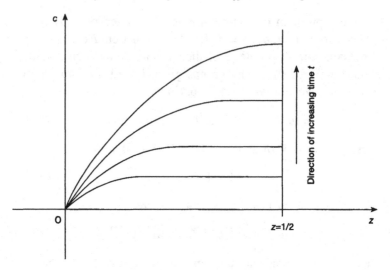

Figure 4.7.1. Sketches of the deflection c of the left-half of the membrane for several times $t > 0$.

At $t = 0$, the membrane, which is initially taut but unstrained, is exposed on one side to a sudden, uniform pressure rise p. Also shown in figure 4.7.1 are sketches of the deflection c of the membrane for several times $t > 0$. For very short times, the central part of the membrane accelerates as though unconstrained; there $c = pt^2/2$. At the edges $z = 0$ and $z = 1$, the membrane is clamped and $c = 0$. Since Eq. (4.7.1) resembles the wave equation, we surmise that for short times a front propagates from the clamped edges towards the center behind which the deflection c is less than $pt^2/2$. If the location of the left-hand front is $z = Z(t)$, then Eq. (4.7.1) becomes

$$c_{tt} = p + c_{zz} \int_0^{Z(t)} c_z^2 \, dz \qquad (4.7.2)$$

which must be solved with the boundary and initial conditions

$$c(z, 0) = 0 \tag{4.7.3a}$$

$$c(0, t) = 0 \tag{4.7.3b}$$

$$c(Z, t) = \frac{pt^2}{2} \tag{4.7.3c}$$

The factor 2 has disappeared from the integral in Eq. (4.7.2) because there are two fronts, one at the left-hand edge and one at the right-hand edge, which make equal contributions to the integral in Eq. (4.7.1).

Equations (4.7.2) and (4.7.3) are invariant to the family of stretching groups

$$c' = \lambda^\delta c \tag{4.7.4a}$$

$$t' = \lambda^{3/2-\delta} t \tag{4.7.4b}$$

$$z' = \lambda z \tag{4.7.4c}$$

$$Z' = \lambda Z \tag{4.7.4d}$$

$$p' = \lambda^{3\delta-3} p \tag{4.7.4e}$$

which embeds the problem posed (for a fixed value of p) in a class of problems of the same kind. The location of the front Z is determined by p and t, which means that $F(Z, p, t) = 0$ for some function $F(Z, p, t)$ that is the same for all problems of the class and thus is an invariant of the family of stretching groups given by Eqs. (4.7.4b, d, e). If we set $\mu = \lambda^{3\delta-3}$, Eqs. (4.7.4b, d, e) can be written as

$$Z' = \mu^\alpha Z \tag{4.7.5a}$$

$$t' = \mu^\beta t \tag{4.7.5b}$$

$$p' = \mu p \tag{4.7.5c}$$

where $\alpha = 1/(3\delta - 3)$ and $\beta = (3/2 - \delta)/(3\delta - 3)$ so that

$$3\alpha - 6\beta = 2 \tag{4.7.6}$$

Thus $M = 3$, $N = -6$ and $L = 2$ so that

$$Z = Ap^{L/M}t^{-N/M} = Ap^{2/3}t^2 \qquad (4.7.7)$$

where A is a constant yet to be determined.

The integral $I = \int_0^{Z(t)} c_z^2 \, dz$ is likewise determined only by p and t through an invariant functional relationship $G(I, p, t) = 0$. Now it follows from Eqs. (4.7.4a, c, d) that $I' = \lambda^{2\delta - 1} I$. Thus G is invariant to

$$I' = \mu^\alpha I \qquad (4.7.8a)$$

$$t' = \mu^\beta t \qquad (4.7.8b)$$

$$p' = \mu p \qquad (4.7.8c)$$

where again $\mu = \lambda^{3\delta - 3}$ but now $\alpha = (2\delta - 1)/(3\delta - 3)$ and $\beta = (3/2 - \delta)/(3\delta - 3)$ so that

$$3\alpha - 6\beta = 4 \qquad (4.7.9)$$

Thus

$$I = Bp^{L/M}t^{-N/M} = Bp^{4/3}t^2 \qquad (4.7.10)$$

where B is another constant. Since I plays the role of the squared velocity in the wave-like equation (4.7.2), we set $I = (dZ/dt)^2$, which yields $B = 4A^2$. Thus Eq. (4.7.2) can be rewritten as

$$c_{tt} = p + (4A^2 p^{4/3})t^2 c_{zz} \qquad (4.7.11)$$

Now if we consider one particular problem of the class, i.e. if we hold p fixed, then Eq. (4.7.2) or (4.7.11) and Eq. (4.7.3) are invariant only to the single group

$$c' = \lambda c \qquad (4.7.12a)$$

$$t' = \lambda^{1/2} t \qquad (4.7.12b)$$

$$z' = \lambda z \qquad (4.7.12c)$$

which is the $\delta = 1$ group of the family (4.7.4). The similarity solution then has the form

$$c = t^2 y\left(\frac{z}{t^2}\right) \qquad (4.7.13)$$

If we substitute Eq. (4.7.13) into Eq. (4.7.11) we obtain

$$2(X^2 - x^2)\ddot{y} + x\dot{y} - y = -\frac{p}{2} \qquad (4.7.14)$$

where $x = z/t^2$ and $X = Ap^{2/3} = Z/t^2$ in view of Eq. (4.7.7). The boundary and initial conditions (4.7.3) become

$$y(0) = 0 \qquad (4.7.15a)$$

$$y(X) = \frac{p}{2} \qquad (4.7.15b)$$

Now Eq. (4.7.14) is an inhomogeneous, linear, ordinary differential equation of the second order. A special solution is $y = p/2$. To this we must add the general solution of the homogeneous equation, one solution of which is $y = x$. The second solution of the homogeneous equation can be found by making d'Alembert's substitution $y = xv$, where v is an auxiliary unknown. After a lengthy but straightforward computation (involving integration by parts), we find

$$y = \frac{p}{2}\left[1 - (X^2 - x^2)^{1/4}X^{-1/2} + \frac{xX^{-1/2}}{2}\int_x^X (X^2 - u^2)^{-3/4}\,\mathrm{d}u\right]$$

$$(4.7.16)$$

as the solution obeying the boundary conditions (4.7.15).

If we insert the similarity solution (4.7.13) into Eq. (4.7.10) we find that

$$\int_0^X \dot{y}^2\,\mathrm{d}x = 4X^2 \qquad (4.7.17)$$

Substituting solution (4.7.16) into Eq. (4.7.17) yields after another lengthy but straightforward computation (again involving integration by parts) the result

$$A = \frac{1}{4}\left\{\left[\frac{2\pi^{1/2}\Gamma(1/4)}{\Gamma(3/4)}\right] - 2\pi\right\}^{1/3} = 0.403\,518\,5970\ldots$$

$$(4.7.18)$$

The solution just obtained applies until the waves propagating in from each clamped edge of the membrane reach the middle. As a last word, let me point out that Eq. (4.7.7), which gives the trajectory of the wave front, has been derived up to an undetermined constant without any detailed computation purely by group-theoretic arguments.

4.8 Further Use of the Associated Group

The invariance of the principal differential equation to the associated group (4.3.1) has other ramifications besides determining the asymptotic behavior in some cases. It follows from Eqs. (4.3.1) that the ratio

$$\frac{\dot{y}(0)}{[y(0)]^{1-M/L}} \equiv C \qquad (4.8.1)$$

being invariant under transformations (4.3.1), is the same for all similarity solutions belonging to the same values of α and β. Remember, the actual function $y(x)$ is different for different values of α and β because the form of the principal differential equation changes as α and β change; compare, for example, Eq. (4.2.7). Thus $C = C(\alpha, \beta)$.

It follows from Eqs. (4.8.1) and (4.2.6) that

$$c_z(0, t) = C(\alpha, \beta)t^{-N/L}[c(0, t)]^{1-M/L} \qquad (4.8.2)$$

In the case of the clamped-flux and clamped-temperature problems of the superfluid diffusion equation (cf. the end of section 4.2), Eq. (4.8.2) can be interpreted as a relation between the heat flux $q(0, t) \equiv -[c_z(0, t)]^{1/3}$ at the front face and the temperature $c(0, t)$ there. Because $M = 2$, $N = -3$ and $L = -4$ for the superfluid diffusion equation,

$$q(0, t) = -[C(\alpha, \beta)]^{1/3}t^{-1/4}[c(0, t)]^{1/2} \qquad (4.8.3)$$

In the clamped-temperature problem for which $\alpha = 0$ and which is analytically integrable, $-[C(\alpha, \beta)]^{1/3} = (3\sqrt{3}/8)^{1/6} =$

$0.9306\ldots$, whereas in the clamped-flux problem for which $\alpha = -1$ and which is not analytically integrable, $-[C(\alpha, \beta)]^{1/3} = 1.095$. In the clamped-temperature problem in which $c(0, t)$ is independent of time, $q(0, t) \sim t^{-1/4}$; in the clamped-flux problem for which $q(0, t)$ is independent of time, $c(0, t) \sim t^{1/2}$.

We can extend the idea just illustrated. Suppose, for example, we wish to know the width of the temperature distribution in the problems just discussed as measured by the average

$$\langle z \rangle \equiv \frac{\int_0^\infty z c\, dz}{\int_0^\infty c\, dz} \tag{4.8.4a}$$

From Eq. (4.2.6) it follows that

$$\langle z \rangle = t^{1/\beta} \frac{\int_0^\infty x y\, dx}{\int_0^\infty y\, dx} \tag{4.8.4b}$$

Now according to Eq. (4.3.1), the quotient of the two integrals in Eq. (4.8.4b) (call it R) transforms according to $R' = \lambda R$ so that

$$\frac{R'}{[y'(0)]^{M/L}} = \frac{R}{[y(0)]^{M/L}} \equiv C_1(\alpha, \beta) \tag{4.8.5}$$

is the same for all similarity solutions belonging to the same value of α and β. Thus

$$\langle z \rangle = C_1(\alpha, \beta) t^{N/L} [c(0, t)]^{M/L} \tag{4.8.6a}$$

which in the case of the superfluid diffusion equation is

$$\langle z \rangle = C_1(\alpha, \beta) t^{3/4} [c(0, t)]^{-1/2} \tag{4.8.6b}$$

Combining Eq. (4.8.6b) with our earlier results, we find that $\langle z \rangle \sim t^{3/4}$ in the clamped-temperature problem and $\langle z \rangle \sim t^{1/2}$ in the clamped-flux problem. One consequence of Eq. (4.8.6b) that might at first appear puzzling is that in the clamped-temperature problem, $\langle z \rangle$ is smaller at any fixed time, the larger the value at which $c(0, t)$ is clamped. This paradoxical result can be explained by remembering that the solution $c_1(z, t)$ for the larger value of $c(0, t)$ and the solution $c_2(z, t)$ for the smaller value of

$c(0, t)$ both have the same asymptotic behavior $[4/(3\sqrt{3})]z^{-2}t^{3/2}$ (see figure 4.8.1). From this figure it is clear that curve 1 weights small values of z more heavily than curve 2.

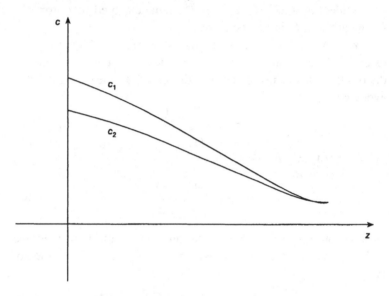

Figure 4.8.1. An auxiliary diagram showing that curve 1 weights small values of z more heavily than curve 2.

4.9   More Wave Propagation Problems

The homogeneous wave equation $c_{tt} = V^2 c_{zz}$ can be separated into

$$u_t = v_z \tag{4.9.1a}$$

$$v_t = V^2 u_z \tag{4.9.1b}$$

if we set $u = c_z$ and $v = c_t$. In the homogeneous form of Eq. (4.7.11), V turned out to be proportional to the time t, and this proportionality to a power of t caused it and Eqs. (4.9.1) to have similarity solutions. If V is a power of u, Eqs. (4.9.1a, *b*) also have similarity solutions, to which physical meaning can be ascribed by interpreting Eqs. (4.9.1*a*, *b*) as describing the stretching of a long, elastic wire [Dr-83]: *v* is the flow velocity of the wire elements, *u* is the strain in the wire, *z* is the Lagrangian coordinate of the wire elements and *t* is the time. The squared velocity $V^2 \equiv (d\sigma/du)/\rho$, where ρ is the initial density of the wire material and $\sigma(u)$ is the strain-dependent tensile stress in the wire. If the near end of the wire is suddenly pulled, a wave of elastic disturbance runs down the wire, and before it reaches the far end of the wire, its progress may be described by a similarity solution (if the near end of the wire is pulled in the right way).

Let σ, the stress in the wire, vary as a power *k* of the strain *u*, i.e. let $\sigma = \rho\mathcal{U}^2 u^k$, where $\mathcal{U}$ is a fiducial velocity that is a material property and appears so that σ will have the correct physical dimensions (remember, strain is dimensionless). Then Eqs. (4.9.1) are invariant to the one-parameter family of groups

$$v' = \lambda^\alpha v \qquad (4.9.2a)$$

$$u' = \lambda^\delta u \qquad (4.9.2b)$$

$$t' = \lambda^\beta t \qquad (4.9.2c)$$

$$z' = \lambda z \qquad (4.9.2d)$$

with the family parameters α, β and δ subject to the constraints

$$(k-1)\alpha + (k+1)\beta = k+1 \qquad (4.9.3a)$$

$$(k-1)\delta + 2\beta = 2 \qquad (4.9.3b)$$

The infinitesimal transformation X of a group G of the family (4.9.2) labeled by the family parameters α, β and δ is

$$X = \alpha v \frac{\partial}{\partial v} + \delta u \frac{\partial}{\partial u} + \beta t \frac{\partial}{\partial t} + z \frac{\partial}{\partial z} \qquad (4.9.4)$$

The invariants of X are

$$x = z/t^{1/\beta}, \quad g = v/t^{\alpha/\beta} \quad \text{and} \quad h = u/t^{\delta/\beta} \qquad (4.9.5)$$

so that similarity solutions take the form $g = g(x)$ and $h = h(x)$, where *g* and *h* are functions yet to be determined.

When $u = t^{\delta/\beta}h(z/t^{1/\beta})$ and $v = t^{\alpha/\beta}g(z/t^{1/\beta})$ are substituted into Eqs. (4.9.1), the following coupled, first-order, ordinary principal differential equations are obtained:

$$\beta\frac{dg}{dx} + x\frac{dh}{dx} = \delta h \qquad (4.9.6a)$$

$$x\frac{dg}{dx} + \beta k\mathcal{U}^2 h^{k-1}\frac{dh}{dx} = \alpha g \qquad (4.9.6b)$$

If we write the subsidiary constraints (4.9.3) in the form

$$M\alpha + N\beta = L \qquad (4.9.7a)$$

$$M'\delta + N'\beta = L' \qquad (4.9.7b)$$

then we can prove with the technique of section 4.3 that the principal differential equations for the functions g and h are invariant to the associated group

$$g' = \mu^{L/M}g \qquad (4.9.8a)$$

$$h' = \mu^{L'/M'}h \qquad (4.9.8b)$$

$$x' = \mu x \qquad (4.9.8c)$$

If X_* is another infinitesimal transformation of the form (4.9.4) belonging to family parameters $\alpha_* \neq \alpha$, $\beta_* \neq \beta$ and $\delta_* \neq \delta$, then X_* and X commute. In (x, g, h)-space,

$$X_* = \left(\frac{\beta - \beta_*}{\beta}\right)\left(x\frac{\partial}{\partial x} + \frac{L}{M}g\frac{\partial}{\partial g} + \frac{L'}{M'}h\frac{\partial}{\partial h}\right) \qquad (4.9.9)$$

which generates the associated group (4.9.8).

The proof of Lie's reduction theorem given in section 3.2 can be used with only minor changes to show that if we introduce the invariants $p = g/x^{L/M}$ and $q = h/x^{L'/M'}$ of the associated group (4.9.8) into the coupled principal equations (4.9.6) for g and h, these equations then reduce to a single, first-order associated differential equation in p and q. In the case of Eqs. (4.9.6), the resulting associated equation is too complex for further analysis, and to continue our discussion, we specialize on a specific problem in the next section.

★4.10   Wave Propagation Problems (continued)

First let us take $k = 1/2$; the stress σ is then an increasing, concave-downward function of the strain u, a realistic behavior. Let us further assume that the wire is long and fastened above to a rigid ceiling. At its lower end ($z = 0$), a large weight is fastened. At $t = 0$, the weight is dropped so that thereafter $v(0, t) = -g_0 t$, where g_0 is the acceleration of gravity. This boundary condition requires that $\alpha/\beta = 1$. Thus $\alpha = \beta = 3/2$ and $\delta = 2$. The other boundary and initial conditions are $v(\infty, t) = u(\infty, t) = 0$ and $v(z, 0) = u(z, 0) = 0$, which require the similarity solutions to obey the boundary condition $g(\infty) = h(\infty) = 0$. Finally, for simplicity, let us choose a system of special units in which $g_0 = \mathcal{U} = 1$. Then

$$x\frac{dp}{dx} = \frac{27pq^{-1/2} - 36p + 12q^{1/2}}{9q^{-1/2} - 8} \equiv G(p, q) \qquad (4.10.1a)$$

$$x\frac{dq}{dx} = \frac{36q^{1/2} - 48q + 18p}{9q^{-1/2} - 8} \equiv F(p, q) \qquad (4.10.1b)$$

where now $p = x^3 g$ and $q = x^4 h$. Division of Eq. (4.10.1a) by (4.10.1b) gives the following associated differential equation in p and q:

$$\frac{dp}{dq} = \frac{27pq^{-1/2} - 36p + 12q^{1/2}}{36q^{1/2} - 48q + 18p} \qquad (4.10.1c)$$

Figure 4.10.1 shows the direction field of Eq. (4.10.1c) in the fourth quadrant of the (q, p)-plane. (Since $v < 0$ and $u > 0$, it is the fourth quadrant that interests us.) The singularities of the right-hand side of Eq. (4.10.1c) are O: $(0, 0)$ and P: $(1/4, -1/3)$. When $x = 0$, $p = q = 0$, so O corresponds to the value $x = 0$. The rule of section 3.4 given on page 49 shows that the singularity P corresponds to the value $x = \infty$. Therefore, for large x, $g \sim p_P/x^3 = -x^{-3}/3$ and $h \sim q_P/x^4 = x^{-4}/4$. Inserting these asymptotic forms into Eqs. (4.9.5) we find

$$v \sim -\frac{t^3}{3z^3} \quad \text{and} \quad u \sim \frac{t^4}{4z^4} \quad \text{when} \quad x = \frac{z}{t^{2/3}} \gg 1 \qquad (4.10.2)$$

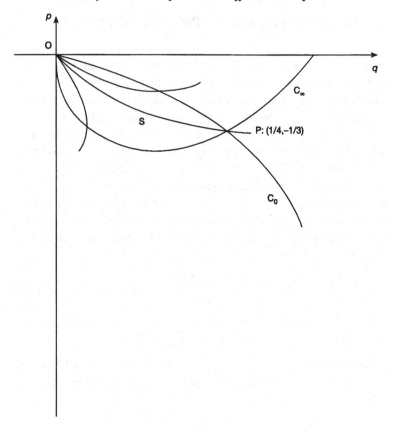

Figure 4.10.1. The fourth quadrant of the direction field of differential equation (4.10.1c). C_0 is the locus of zero slope and C_∞ is the locus of infinite slope.

The asymptotic forms (4.10.2) are exactly the exceptional (totally invariant) solutions $v = Az^{L/M}t^{-N/M}$ and $u = A'z^{L'/M'}t^{-N'/M'}$. We could have showed that these exceptional solutions describe the asymptotic behavior of solutions of the wave equations (4.9.1) by adapting the work of sections 4.4 and 3.3. This adaptation differs from the earlier work only in the proof of the ordering theorem, which takes a different line for wave propagation equations than it does for diffusion equations.

To show the ordering, we use Riemann's *method of characteristics*† [Co-48]. If we multiply Eq. (4.9.1a) by $\pm V$ and add it to Eq. (4.9.1b), we find after some minor rearrangement that

$$\left(v \pm \int\limits_0^u V \, du \right)_t - (\pm V)\left(v \pm \int\limits_0^u V \, du \right)_z = 0 \qquad (4.10.3)$$

Equation (4.10.3) means that the quantities $v \pm \int_0^u V \, du$ are constant along the curves $dz \pm V \, dt = 0$. The quantities $v \pm \int_0^u V \, du$ are called *Riemann invariants* and the curves $dz \pm V \, dt = 0$ are called *characteristics*.

Figure 4.10.2 shows the first quadrant of the (z, t)-plane with the boundary and initial conditions $v(\infty, t) = u(\infty, t) = 0$ and $v(z, 0) = u(z, 0) = 0$ explicitly displayed. On every negative characteristic $dz = -V \, dt$, for example, characteristic PQ, $v + \int_0^u V \, du = 0$. Since a negative characteristic joins every point in the first quadrant with a point on the z-axis, at every such point $v = -\int_0^u V \, du$. On every positive characteristic $dz = V \, dt$, therefore, the quantity $v - \int_0^u V \, du = -2\int_0^u V \, du$ is conserved. Since V is a function only of u, this means that on positive characteristics u, v and V are constant. The positive characteristics are therefore straight lines.

In the problem at hand (in the special units in which $g_0 = \mathcal{U} = 1$), $\sigma = \rho u^{1/2}$, so that $V = (4u)^{-1/4}$. Since $v(0, t) = -t$, $u(0, t) = (3t/2\sqrt{2})^{4/3}$ and $V(0, t) = (3t)^{-1/3}$. Thus $V(0, t)$ is a decreasing function of t. This means that the positive characteristic lines, whose equations are $z = V(0, t_0)(t - t_0)$, spread apart fanwise as shown in figure 4.10.2.

Consider now two solutions (u_1, v_1) and (u_2, v_2) such that $V_1(0, t)$ and $V_2(0, t)$ are decreasing functions of time for which $V_1(0, t) < V_2(0, t)$. We prove $V_1(z, t) < V_2(z, t)$ by supposing the opposite, namely $V_1(z, t) > V_2(z, t)$, as at point R in

† The reader unfamiliar with Riemann's method of characteristics will find a brief introduction in Appendix B.

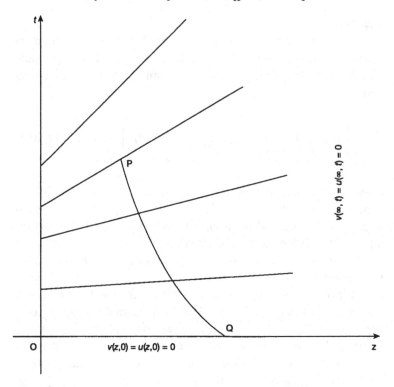

Figure 4.10.2. The first quadrant of the (z, t)-plane showing a negative characteristic and several positive characteristics.

figure 4.10.3, where the lines AB and A′B′ are the positive characteristics of solutions 1 and 2 that pass through the point R. Then $V_2(A') = V_2(R) < V_1(R) = V_1(A) < V_2(A)$ which contradicts the hypothesis that $V_2(0, t)$ is a decreasing function of time. Thus the values of V are ordered according to their values $V(0, t)$ at the origin. As long as V is a monotone function of u (either decreasing or increasing), both u and v are ordered according to their values $u(0, t)$ and $v(0, t)$ at the origin.

The Riemann method can provide a direct route to the asymptotic forms of u and v. When t is sufficiently large

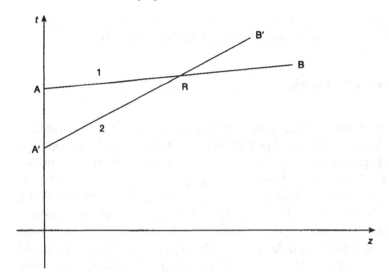

Figure 4.10.3. An auxiliary diagram showing the first quadrant of the (z, t)-plane used in the proof of the ordering.

compared with t_0, $V(0, t_0) \sim z/t$. Now since $V(0, t_0)$ is the value of V on the entire characteristic, if $V = f(u)$ is the connection between V and u, then

$$u \sim f^{-1}\left(\frac{z}{t}\right) \tag{4.10.4}$$

In the problem at hand, $V = (4u)^{-1/4}$, so that Eq. (4.10.4) becomes Eq. (4.10.2). More generally,

$$u \sim \left(\frac{z}{t\sqrt{k}}\right)^{2/(k-1)} \tag{4.10.5a}$$

$$v \sim -\left(\frac{2\sqrt{k}}{k+1}\right)\left(\frac{z}{t\sqrt{k}}\right)^{(k+1)/(k-1)} \tag{4.10.5b}$$

Equations (4.10.5) can only obey the conditions $v(\infty, t) = u(\infty, t) = 0$ and $v(z, 0) = u(z, 0) = 0$ if $k < 1$. This also follows from the form of the exceptional solutions $v =$

$Az^{L/M}t^{-N/M}$ and $u = A'z^{L'/M'}t^{-N'/M'}$ since $L/M = N/M = (k+1)/(k-1)$ and $L'/M' = N'/M' = 2/(k-1)$.

★4.11   Shocks

In figure 4.10.2, corresponding to $k = 1/2$, the positive characteristics diverged fanwise. But for $k > 1$, the positive characteristics converge. For then (still in special units), $V = (\sqrt{k})u^{(k-1)/2}$ and $v = -[2\sqrt{k}/(k+1)]u^{(k+1)/2}$ so that if $v(0, t) = -t$, $V(0, t) = \sqrt{k}[(k+1)t/(2\sqrt{k})]^{(k-1)/(k+1)}$. When $k < 1$, $V(0, t)$ is a decreasing function of t and the positive characteristics diverge; when $k > 1$, $V(0, t)$ is an increasing function of t and the positive characteristics converge. It should be noted that this conclusion depends on the boundary condition $v(0, t) = -t$.

Now, as is well known, two positive characteristics cannot intersect, for if they did, the two Riemann invariants would be over-determined by the two positive and one negative characteristic through the point of intersection. Instead, a shock front forms across which u and v jump discontinuously and to either side of which only two characteristics pass through each point.

The velocity U of the shock front is connected to the magnitudes of the jumps in u and v, and the connection depends on the differential equations (4.9.1a, b). Both of these equations are in the form of conservation equations, the generic form of which is Eq. (4.5.1), but with $S(c) = 1$. Here c is best thought of as a concentration (having, for example, the dimensions of moles m^{-3}) and q as its flux (having dimensions moles m^{-2} s^{-1}). Suppose c and q have the respective values c_1 and q_1 to the left and c_2 and q_2 to the right of a shock front propagating from left to right with velocity U. In a frame of reference stationary with respect to the shock (i.e. moving to the right with velocity U), material with concentration c_2 appears to be approaching the shock front with a velocity $-U$ and material with concentration c_1 appears to be receding from it with the same velocity. Conservation of material at the shock front requires that

$$q_2 - Uc_2 = q_1 - Uc_1 \qquad (4.11.1a)$$

The expression on either side is the total current of material in the shock-stationary frame. Thus

$$U = \frac{\Delta q}{\Delta c} \qquad (4.11.1b)$$

where the jumps Δq and Δc are taken in the same direction.

If we consider solutions for which $v = u = 0$ before the shock front, this being the way in which the conditions are fulfilled $v(\infty, t) = u(\infty, t) = 0$, then for Eqs. (4.9.1a, b) the shock relations are

$$U = -\frac{v}{u} \qquad (4.11.2a)$$

$$U = -\frac{\int_0^u V^2 \, du}{v} \qquad (4.11.2b)$$

where v and u are the values just behind the shock front. Since $V^2 = ku^{k-1}$, Eqs. (4.11.2a, b) become

$$U = -\frac{v}{u} = -\frac{u^k}{v} \qquad (4.11.2c)$$

If the conservation equation $c_t + q_z = 0$ is invariant to the family of groups $c' = \lambda^\varepsilon c$, $q' = \lambda^\varphi q$, $t' = \lambda^\kappa t$, $z' = \lambda z$, then according to Eq. (4.11.1b), $U' = \lambda^{\varphi-\varepsilon}U$. But the invariance of the conservation equation requires ε, φ and κ to obey the constraint $\varepsilon - \kappa = \varphi - 1$. Thus $U' = \lambda^{1-\kappa}U$, just as a velocity should. Conversely, if we take $U' = \lambda^{1-\kappa}U$ because U is a velocity, then the jump relation (4.11.1b) is invariant to the same group as the conservation equation.

For similarity solutions of the form given by Eqs. (4.9.5), the shock front has the trajectory $z = Xt^{1/\beta}$, where X is a constant yet to be determined. Then $U = (X/\beta)t^{(1-\beta)/\beta}$ and the jump conditions (4.11.2c) become

$$\frac{X}{\beta} = -\frac{g(X)}{h(X)} = -\frac{[h(X)]^k}{g(X)} \qquad (4.11.3)$$

If we choose a value of X, we can calculate values of $g(X)$ and $h(X)$ from Eqs. (4.11.3) and so have sufficient initial conditions for a backwards numerical integration of the principal differential equations (4.9.6). If the value of g or h at the origin so obtained is not correct, we can scale the solution using the associated group (4.9.8) to make it correct. Thus only a single integration need be undertaken.

Notes

Note 1: The mean value theorem for integrals is used to evaluate $S(c_1) c_{1t} - S(c_2) c_{2t}$ as follows:

$$S(c_1) c_{1t} - S(c_2) c_{2t} = \frac{\partial}{\partial t} \int_{c_1}^{c_2} S(c)\, dc = \frac{\partial}{\partial t} \underline{S} u \qquad (4.11.4)$$

where the underlined quantity $\underline{S}$ stands for $S(c)$ evaluated at some undetermined value intermediate between c_1 and c_2 and $u = c_1 - c_2$. Furthermore, according to the chain rule

$$\frac{\partial}{\partial t} \underline{S} u = \underline{S}_t u + \underline{S} u_t = \underline{S}_c c_t u + \underline{S} u_t \qquad (4.11.5)$$

which is the form needed for the derivation of two of the terms in Eq. (4.5.2).

The mean value theorem is used to evaluate $q_z(z, t, c_1, c_{1z}) - q_z(z, t, c_2, c_{2z})$ as follows:

$$q_z(z, t, c_1, c_{1z}) - q_z(z, t, c_2, c_{2z}) = \frac{\partial}{\partial z}[q(z, t, c_1, c_{1z}) - q(z, t, c_2, c_{2z})]$$

$$= \frac{\partial}{\partial z}\left(\frac{\partial q}{\partial c} u - \frac{\partial q}{\partial c_z} u_z \right)$$

where again the underlined quantities are evaluated at some undetermined value intermediate between c_1 and c_2 and $u = c_1 - c_2$. The indicated z-differentiation yields the remainder of the required terms in Eq. (4.5.2).

Problems for Chapter 4

4.1 An engineer conducts heat transfer experiments in a long rod (taken as semi-infinite) by measuring the temperature rise $c(z, t)$

at various points z on the rod as a function of the elapsed time t. In each experiment, he increases the temperature rise $c(0, t)$ at one end of the rod proportionally to a power of the time, these powers being different in different experiments. When he plots his data, he notices that far from the heated end the temperature rise in all experiments falls as the same power a of z at each fixed time and rises with the same power b of t at each fixed position. He guesses that the temperature rise is described by a similarity solution like Eq. (4.2.6). To check his guess, he deduces from his data taken far from the heated end that the temperature $c(0, t)$ at the heated end and its space derivative $c_z(0, t)$ should obey a relation $[c(0, t)]^k c_z(0, t) \sim t^m$ that is the same for all his experiments. What are the exponents k and m in this relation in terms of the measured exponents a and b?

4.2 Determine the coefficients M, N and L in the linear constraint equation (4.1.2) for the illustrative partial differential equation (3): $cc_t = c_{zz}$ (see section 4.1). Find the coefficient A of the exceptional solution $c = Az^{L/M} t^{-N/M}$ by direct substitution. Now substitute the similarity form (4.2.6) into the partial differential equation and determine the form of the principal ordinary differential equation. Check to see that it is invariant to the associated group (4.3.1a, b). Choose the invariant $u = y/x^{L/M}$ and the first differential invariant $v = \dot{y}/x^{L/M-1}$ and use Lie's reduction theorem to reduce the order of the principal differential equation. Find the singularities of the reduced equation and so determine the coefficient A in the exceptional solution in a second way. When $\alpha = -1/2$, the principal differential equation is easily solvable for the family of solutions that vanish at infinity. Find this family. Does it have the asymptotic form expected from the exceptional solution? What curve in the (u, v)-plane does the family of solutions correspond to? Show that this curve is a solution of the reduced differential equation.

4.3 Suppose that for a particular problem the exponents α and β in the transformation equations (4.1.1a–c) have the fixed values $\alpha = \alpha°$, $\beta = \beta°$. If we transform the similarity solution

$c = t^{\alpha^\circ/\beta^\circ} y(z/t^{1/\beta^\circ})$ with the group of transformations of the family corresponding to $\alpha = \alpha^\circ$, $\beta = \beta^\circ$ it remains unchanged. What happens if we transform it with a group of transformations of the family for which $\alpha \neq \alpha^\circ$, $\beta \neq \beta^\circ$? (Remember that $M\alpha + N\beta = L$ and $M\alpha^\circ + N\beta^\circ = L$.) Can you use the resulting expression to prove the existence of the associated group (4.3.1)?

4.4   Pattle [Pa-59] has studied the diffusion equation $c_t = (\mathcal{D}(c)\, c_z)_z$ when the concentration-dependent diffusion constant $\mathcal{D}(c) = c$. In this special case, the partial differential equation is invariant to the family of stretching groups (4.1.1a–c) with $M = 1$, $N = 1$ and $L = 2$ as the coefficients in the linear constraint (4.1.2). Determine the principal ordinary differential equation obeyed by the function $y(x)$, $x \equiv z/t^{1/\beta}$, in the similarity solution (4.2.6). Using the invariants $u = y/x^{L/M}$, $v = \dot{y}/x^{L/M-1}$ of the associated group and Lie's reduction theorem, reduce the principal ordinary differential equation to first order.

When $\alpha = -1$, the principal equation is easily solvable for a family of positive, decreasing solutions that vanish at infinity. Solve it. To what curve C in the (u, v)-plane does the family correspond? To what point Q on the curve C does $x = 0$ correspond? To what point P on the curve C does $y = 0$ correspond? When $y = 0$, how are x and $\dot{y}$ related? (Hint: consider the value of v_P.)

Sketch the direction field of the first-order reduced equation when $\alpha = 1/2$ and $\beta = 3/2$. Locate the singularities. Positive, decreasing solutions $y(x)$ correspond to curves in the fourth quadrant of the (u, v)-plane. Is there a singularity in that quadrant from which you can determine simultaneously a value of x, y and $\dot{y}$? Could these values serve as a starting point for a numerical integration of the principal equation? What values would y have for values of x larger than the starting value? How many numerical integrations would you have to carry out to determine the entire family of positive, decreasing solutions that vanish at infinity?

4.5   The pulsed-source problem in an infinite medium for the nonlinear diffusion equation $c_t = (c^n c_z)_z$ is defined by the boundary, initial, and conservation conditions $c(\pm\infty, t) = 0$, $t > 0$; $c(z, 0) = 0, |z| > 0$; $\int_{-\infty}^{+\infty} c(z, t) \, dz = 1$, $t > 0$. Find an explicit formula for the solution $c(z, t)$.

4.6 The linear diffusion equation $c_t = c_{zz}$ is invariant to the one-parameter family of stretching groups $c' = \lambda^\alpha c$, $t' = \lambda^2 t$, $z' = \lambda z$, where α can have any value. This family has the form (4.1.1) with the coefficients $M = 0$, $N = 1$, $L = 2$ in the linear constraint (4.1.2). When $M = 0$, the equation (4.3.1) for the associated group involves an inadmissible division by zero. Can you develop a simple, heuristic argument that enables you to guess the actual form of the associated group? Check your guess for the principal ordinary differential equation of the linear diffusion equation. Now adapt the reasoning of section 4.3 to prove your guess. Finally, use Lie's theorem and reduce the order of the principal differential equation.

4.7 Sometimes it is easier to solve the principal differential equation directly than to use Lie's theorem to reduce its order. The principal differential equation $2\ddot{y} + x\dot{y} - \alpha y = 0$ of the linear diffusion equation of problem 4.6 is a good example. It is easily solved when $\alpha = 0, -1$ and 1. Find the solutions that vanish at infinity for these values of α. (Hint: when $\alpha = -1$, the *linear* principal differential equation has the particular solution $y = x$. Use the method mentioned in problem 3.5 to complete the solution.)

4.8 Does the ordering theorem of section 4.5 hold if we replace the condition $c_1(0, t) \geq c_2(0, t)$ for $0 < t < T$ by the condition $c_{1z}(0, t) < c_{2z}(0, t)$ for $0 < t < T$?

4.9 Barenblatt and Zeldovich [Ba-72] studied the nonlinear diffusion equation $c_t = (cc_z)_z$ in a half-space with the boundary condition $c(0, t) = e^t$. They noted that the partial differential equation is invariant to the one-parameter family of one-parameter groups $c' = \lambda^2 c$, $t' = t + \beta \ln \lambda$, $z' = \lambda z$. Write an expression for the similarity solutions and using the approach of section 4.3 find the associated group. Determine the principal ordinary differential equation and verify that it is invariant to the associated group that you have found. What value of β does the boundary condition $c(0, t) = e^t$ determine?

4.10 Find a family of groups that leaves invariant the partial differential equation $c_t = (e^c c_z)_z$. Write an expression for the similarity solution. Use the reasoning of section 4.3 to find the associated group. When the boundary and initial conditions are $c(0, t) = a, t > 0$; $c(\infty, t) = b, t > 0$; $c(z, 0) = b, z > 0$, what form does the similarity solution take? If y is the dependent variable in the principal ordinary differential equation, use the associated group to show that $a - b$ is a function of $\dot{y}^2(0)\, e^a$.

4.11 In the struck-membrane problem of section 4.7, use the method outlined in Eqs. (4.7.4)–(4.7.7) to find the dependence of $c_z(0, t)$ on p and t. Check your answer by referring to Eqs. (4.7.13) and (4.7.16).

4.12 The motion of water in a long, narrow, open channel is described by the coupled partial differential equations

$$h_t + (vh)_z = 0 \qquad (1a)$$
$$v_t + vv_z + h_z = 0 \qquad (1b)$$

where z is the longitudinal position coordinate, t is the time, h is the height of the water above the channel floor and v is the longitudinal flow velocity.

(a) Find a one-parameter family of groups like that in Eqs. (4.9.2a–d) to which Eqs. (1a, b) are invariant.
(b) Write expressions for the similarity solutions.
(c) Consider the solution $h(z, t)$, $v(z, t)$ that corresponds to the initial conditions $v(z, 0) = 0, -\infty < z < \infty$; $h(z, 0) = h_o$, $z > 0$; $h(z, 0) = 0, z < 0$. These initial conditions describe a semi-infinite channel filled on one side with still water of depth h_o and empty on the other side. At time $t = 0$, the restraining dam at $z = 0$ ruptures. The partial differential equations (1) determine the subsequent motion of the water. What form do the similarity solutions now take?

(d) Determine the associated group [see Eqs. (4.9.8a–d)]. Determine from the associated group the form of the dependence on h_o of the coordinate $z_o > 0$ of the front separating the water set in motion by the breaking of the dam from the undisturbed water upstream of the dam (where v is still 0 and h is still h_o).

(e) Find explicit expressions for the similarity solution. How far downstream ($z < 0$) does the disturbance extend?

5

Traveling-Wave Solutions

5.1 One-Parameter Families of Translation Groups

Chapter 4 is devoted to exploring similarity solutions of partial differential equations invariant to the family (4.1.1) of stretching groups. As the reader will recall, the similarity solutions are those which are invariant to *one* group of the family (4.1.1). The present chapter is similarly devoted to exploring certain solutions of partial differential equations invariant to the following family of translation groups:

$$
\boxed{
\begin{aligned}
c' &= c \\
t' &= t + \lambda \qquad -\infty < \lambda < \infty \\
z' &= z + \alpha\lambda
\end{aligned}
}
\qquad
\begin{aligned}
&(5.1.1a) \\
&(5.1.1b) \\
&(5.1.1c)
\end{aligned}
$$

Here again λ is the group parameter that labels different transformations of a group and α is the family parameter that labels different groups of the family. As in chapter 4, the solutions we shall study are those that are invariant to *one* group of the family. For reasons noted below, solutions invariant to one group of the family (5.1.1) are not called similarity solutions but rather *traveling-wave solutions*.

Because of the complexity that arises in the study of traveling-wave solutions, we restrict our attention in this chapter to a single, important equation invariant to the family (5.1.1), namely, the one-dimensional heat diffusion equation with a temperature-dependent source, Eq. (4.6.1):

$$c_t = c_{zz} + Q(c) \tag{5.1.2}$$

This equation occurs in many applications, among them applied superconductivity (as mentioned in chapter 4), the propagation of epidemics, and the growth and migration of populations. It is the focus of our attention in this chapter because, as the reader will see subsequently, detailed study of its traveling-wave solutions already uncovers many of the problems involved in determining the traveling-wave solutions of any partial differential equation invariant to the family (5.1.1).

A solution $c(z, t)$ of the partial differential equation that is invariant to a group of the family belonging to the family parameter α, obeys the condition

$$c'(z', t') = c(z, t) \tag{5.1.3a}$$

which can be written

$$c(z + \alpha\lambda, t + \lambda) = c(z, t) \tag{5.1.3b}$$

If we differentiate Eq. (5.1.3b) with respect to λ and then set $\lambda = 0$, we obtain the linear partial differential equation

$$\alpha c_z + c_t = 0 \tag{5.1.4}$$

whose characteristic equations are

$$\frac{dz}{\alpha} = \frac{dt}{1} = \frac{dc}{0} \tag{5.1.5}$$

Two independent integrals of Eqs. (5.1.5) are

$$x \equiv z - \alpha t \quad \text{and} \quad c \tag{5.1.6}$$

Thus the most general solution of Eq. (5.1.4) is

$$\boxed{c = y(z - \alpha t)} \tag{5.1.7}$$

where y is an as yet undetermined function of the argument $x \equiv z - \alpha t$.

A solution of the form (5.1.7) is called a *traveling-wave solution*. Such solutions interest us for two reasons. First, being invariant solutions, they are generally easier to find than other solutions. And second, they often represent the late-time asymptotic behavior approached by solutions arising from quite different initial conditions.

If we substitute the form (5.1.7) into the original partial differential equation, we obtain the principal ordinary differential equation for the function $y(x)$. We now prove that this principal differential equation is invariant to the associated group

$$
\begin{array}{ll}
y' = y & \quad (5.1.8a) \\
x' = x + \mu & \quad (5.1.8b)
\end{array}
\qquad -\infty < \mu < \infty
$$

We follow the procedure of section 4.3. The infinitesimal transformation X of a group G of the family (5.1.1) belonging to the family parameter α is

$$
X = \alpha \frac{\partial}{\partial z} + \frac{\partial}{\partial t} \qquad (5.1.9a)
$$

Note that the coefficient of $\partial/\partial c$ in X is zero. Besides the infinitesimal transformation X, there are other infinitesimal transformations that leave the original partial differential equation invariant, namely,

$$
X_* = \alpha_* \frac{\partial}{\partial z} + \frac{\partial}{\partial t} \qquad (5.1.9b)
$$

where $\alpha_* \neq \alpha$. The commutator of X and X_* vanishes. Since $Xx = 0$ and $Xy = 0$ by the definition of x and y ($\equiv c$) as invariants of X, it follows that $X(X_* x) = 0$ and $X(X_* y) = 0$. Therefore, $X_* x$ and $X_* y$ are invariants of X and so can be expressed as functions of x and y only. According to the chain rule, then, the action of X_* on any function of x and y is given by

$$
X_* = (X_* x) \frac{\partial}{\partial x} + (X_* y) \frac{\partial}{\partial y} \qquad (5.1.10)
$$

We can calculate the functions $X_* x$ and $X_* y$ explicitly using Eq. (5.1.9b) and the definitions $x \equiv z - \alpha t$ and $y \equiv c$. Thus

$$X_* x = \alpha_* - \alpha \qquad (5.1.11a)$$

$$X_* y = 0 \qquad (5.1.11b)$$

since the coefficient of $\partial/\partial c$ in X_* is also zero. Therefore

$$X_* = (\alpha_* - \alpha)\frac{\partial}{\partial x} \qquad (5.1.11c)$$

The orbits of the group G_* generated by X_* in (x, y)-space are obtained by solving the differential equations

$$\frac{dx}{(\alpha_* - \alpha)} = \frac{dy}{0} = d\mu \qquad (5.1.12)$$

They are Eqs. (5.1.8a, b), as claimed.

Because X_* is an infinitesimal transformation of the group (5.1.1), it must carry one solution of the partial differential equation into another. Therefore, the transformations of G_* in (x, y)-space carry one traveling-wave solution $y(x)$ belonging to the family parameter α into another traveling-wave solution $y'(x')$ belonging to the same family parameter. The totality of traveling-wave solutions belonging to a particular family parameter then form an invariant family, which demonstrates that the principal differential equation is invariant to the associated group (5.1.8) for any choice of the family parameter α.

5.2 The Diffusion Equation with Source

If we substitute the form (5.1.7) into the partial differential equation (5.1.2), we obtain the principal differential equation

$$\ddot{y} + \alpha \dot{y} + Q(y) = 0 \qquad (5.2.1)$$

which is, as expected, invariant to the associated group (5.1.8). The quantity y is an invariant of the associated group and the quantity $u = \dot{y}$ is a first differential invariant. According to

Lie's reduction theorem (section 3.2), if we use y and u for new variables, Eq. (5.1.1) reduces to a first-order associated differential equation, which turns out to be

$$\dot{u} \equiv \frac{du}{dy} = -\alpha - \frac{Q(y)}{u} \qquad (5.2.2)$$

This is as far as we can go in general terms without specifying something about the source function $Q(y)$. A number of different forms have been investigated. In the case of population growth and migration, the function

$$Q(c) = c(1 - c) \qquad (5.2.3)$$

has been used, in which case Eq. (5.1.2) is called the Fisher equation [Mu-89]. In applied superconductivity the form $Q(c) = \gamma W(c) - c$ is quite often used for superconductors cooled with liquid helium [Dr-95]. Here

$$W(c) = \begin{cases} 0 & 0 \le c \le c_0 & (5.2.4a) \\ \dfrac{c - c_0}{1 - c_0} & c_0 \le c \le 1 & (5.2.4b) \\ 1 & 1 \le c & (5.2.4c) \end{cases}$$

and γ is a constant. The form $Q(c) = \gamma W(c)$ is used for uncooled superconductors. Sometimes the following simplified form is used for $W(c)$:

$$\begin{aligned} W(c) &= 0 & 0 \le c < 1 & \qquad (5.2.5a) \\ &= 1 & 1 \le c & \qquad (5.2.5b) \end{aligned}$$

5.3 Determination of the Propagation Velocity α

Since $x \equiv z - \alpha t$, the family parameter α is the velocity at which the traveling wave (5.1.7) propagates in the direction of increasing z. In chapter 4, the family parameters were determined directly from the boundary and initial conditions. In this chapter, too, the boundary conditions determine the family parameter, although, as we shall see, not as straightforwardly as in chapter 4. But first we must decide what boundary conditions to consider.

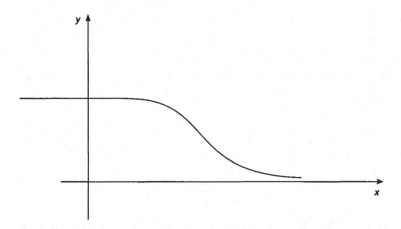

Figure 5.3.1. A traveling-wave shape that is flat far ahead of and far behind the wave front.

We expect the interesting traveling-wave solutions $y(x)$ to look like the curve shown in figure 5.3.1. Far ahead of the wave front and far behind it, $y(x)$ is flat. In those regions where $y(x)$ is flat, $\ddot{y} = \dot{y} = 0$, from which it follows that $c_t = c_{zz} = 0$. Then the partial differential equation (5.1.2) reduces to $Q(c) = 0$. Thus the flat asymptotes of the traveling-wave solution must be roots of the source function $Q(c)$.

Suppose for the sake of argument we choose (for the remainder of section 5.3) $Q(c) = \gamma W(c) - c$ where $W(c)$ is given by Eqs. (5.2.5). If $\gamma < 1$, there is only one root, $c = 0$. If $\gamma > 1$, there are three roots, $c = 0$, $c = 1$ and $c = \gamma$ (see figure 5.3.2). Thus if $\gamma < 1$, there can be no traveling-wave solution of the kind depicted in figure 5.3.1. So let us further assume for the remainder of section 5.3 that $\gamma > 1$.

Of the three flat steady solutions, $c = 0$, $c = 1$ and $c = \gamma$, the first and the third are stable against small perturbations, while

Traveling-Wave Solutions

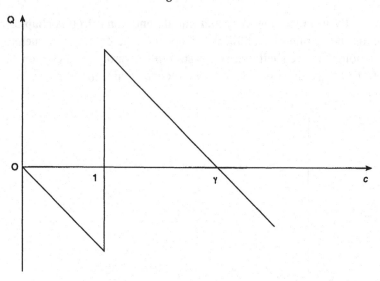

Figure 5.3.2. The function $Q(c) = \gamma W(c) - c$ when $W(c)$ is given by Eq. (5.2.5) and $\gamma > 1$.

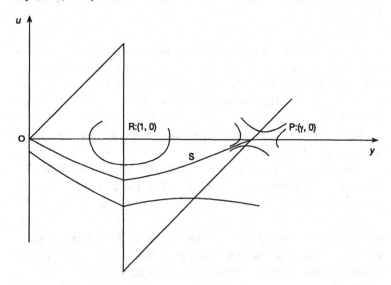

Figure 5.3.3. Part of the direction field of differential equation (5.2.2).

the second is unstable.   We consider here perturbations that
depend on time alone; a complete treatment is given in note 1 at
the end of the chapter. Let $c = \underline{c} + \varepsilon(t)$, where $\underline{c}$ is a root of
$Q(c) = 0$ and $\varepsilon(t) \ll 1$. Then to first order, Eq. (5.1.2) becomes

$$\frac{d\varepsilon}{dt} = \varepsilon \left(\frac{dQ}{dc} \right)_{c=\underline{c}} \tag{5.3.1}$$

An initial perturbation will decay if $(dQ/dc)_{c=\underline{c}} < 0$ ($\underline{c} = 0$ or
$\underline{c} = \gamma$) and grow if $(dQ/dc)_{c=\underline{c}} > 0$ ($c = 1$)†. Since only stable
asymptotes can endure in nature, we are interested in traveling
waves for which $y(-\infty) = \gamma$ and $y(\infty) = 0$.

Figure 5.3.3 shows the right half of the (y, u)-plane. The
vertical line $u = -Q(y)/\alpha$ is the locus of zero slope ($\dot{u} = 0$).
The y-axis ($u = 0$) is the locus of infinite slope ($\dot{u} = \infty$). The
intersections of these loci at O: $(0, 0)$, R: $(1, 0)$ and P: $(\gamma, 0)$ are
the singular points of the differential equation (5.2.2). The loci
of zero and infinite slope divide the (y, u)-plane into regions
in which $\dot{u}$ has one algebraic sign.   Some integral curves of
Eq. (5.2.2) have been sketched in figure 5.3.3. The curves in
each region show the sign of $\dot{u}$ there. *Note that because of the
jump in $Q(y)$ at $y = 1$, $\dot{u}$ is not continuous there.* Because $y > 0$
and $u = \dot{y} \leq 0$ for the traveling wave of figure 5.3.1, the wave
corresponds to the integral curve in the fourth quadrant joining
point O with point P; it is a separatrix S. We can find this integral
curve and the propagation velocity α to which it corresponds as
follows.

When $y < 1$, Eq. (5.2.2) with $Q(c) = \gamma W(c) - c$ and $W(c)$
given by Eq. (5.2.5) becomes

$$\dot{u} = -\alpha + \frac{y}{u} \tag{5.3.2a}$$

The solution with $u = \dot{y} < 0$ that vanishes at the origin is
$u = -\kappa_+ y$, where κ_+ is the positive root of

$$\kappa^2 - \alpha\kappa - 1 = 0 \tag{5.3.2b}$$

---

† This demonstration is sufficient to prove the instability of the second root but not to
prove the stability of the first and third roots, for which the more general procedure of
note 1 is required.

namely

$$\kappa_+ = [\alpha + (\alpha^2 + 4)^{1/2}]/2 \qquad (5.3.2c)$$

as can be shown by direct substitution (see note 2 for further details).

When $\gamma > y > 1$, Eq. (5.2.2) becomes

$$\dot{u} = -\alpha - \frac{(\gamma - y)}{u} \qquad (5.3.3a)$$

The solution with $u = \dot{y} \le 0$ vanishing at $y = \gamma$ is $u = -\kappa_-(y-\gamma)$, where κ_- is the negative root of Eq. (5.3.2b), namely

$$\kappa_- = [\alpha - (\alpha^2 + 4)^{1/2}]/2 \qquad (5.3.3b)$$

The two partial solutions just obtained must be equal at $y = 1$. (Note, however, that because of the jump in $Q(y)$ at $y = 1$, their slopes need not be the same there.) Therefore

$$-\kappa_+ = \kappa_-(\gamma - 1) \qquad (5.3.4a)$$

from which it follows that

$$\alpha = \frac{(\gamma - 2)}{(\gamma - 1)^{1/2}} \qquad (5.3.4b)$$

If $1 < \gamma < 2$, then $\alpha < 0$ and the wave propagates from right to left (state $c = \gamma$ collapses) whereas if $\gamma > 2$, then $\alpha > 0$ and the wave propagates from left to right (state $c = \gamma$ grows). When $\gamma = 2$, $\alpha = 0$.

The value $\gamma = 2$ for $\alpha = 0$ can easily be obtained from Eq. (5.2.2) by writing Eq. (5.2.2) in the form $u\dot{u} = -Q(y)$ when $\alpha = 0$. Then integrate with respect to y from 0 to γ. If we note that $u\dot{u}$ is a perfect differential and that $u(0) = u(\gamma) = 0$, we see that

$$\int_0^\gamma Q(y)\,dy = 0 \qquad (5.3.5)$$

when $\alpha = 0$. This means that the positive and negative lobes of $Q(c)$ in figure 5.3.2 have equal areas from which it follows that

$\gamma = 2$. In applied superconductivity, Eq. (5.3.5) is known as the equal-areas theorem of Maddock, James and Norris [Ma-69] and is applied to source functions $Q(y)$ more complicated than those being studied here.

In applied superconductivity, traveling waves that span the steady, flat states $c = 0$ and $c = \gamma$ are actually created in helium-cooled superconducting wires (here interpret c as the local temperature of the wire). On the other hand, a traveling wave that spans the steady, flat states $c = 0$ and $c = 1$, though it exists mathematically, cannot be maintained in the laboratory because it is unstable. For such a traveling wave, $y = 1$ for $x < a$ and $\dot{y} = u = -\kappa_+ y$ so that $y = \exp[-\kappa_+(x - a)]$ for $x > a$. Similar unstable traveling waves connect the flat, steady states $c = 1$ and $c = \gamma$. For such a traveling wave, $y = 1$ for $x > a$ while for $x < a$, $y = \gamma + (1 - \gamma)\exp[-\kappa_-(x - a)]$. Unstable traveling-wave solutions of the latter two kinds exist for all α.

5.4 Determination of the Propagation Velocity: Role of the Initial Condition

Let us now consider a case in which $Q(c) = \gamma W(c)$ with $W(c)$ still given by Eq. (5.2.5). Now the equation $Q(c) = 0$ has roots for all values of c for which $0 \leq c < 1$. These roots behave as though they are stable in the sense that a small perturbation does not grow, but instead spreads by diffusion, so that its amplitude decreases while its integrated strength remains constant. One can prove by direct calculation that there is no traveling wave of the kind depicted in figure 5.3.1 spanning two flat, steady states each with a value of c in the interval $0 \leq c < 1$. But there are other traveling-wave solutions as we now show.

When $Q(c) = \gamma W(c)$, Eq. (5.2.2) becomes

$$\dot{u} = -\alpha \qquad\qquad y < 1 \qquad\qquad (5.4.1a)$$

$$\dot{u} = -\alpha - \frac{\gamma}{u} \qquad y > 1 \qquad\qquad (5.4.1b)$$

Shown in figure 5.4.1 is the right half of the (y, u)-plane. When

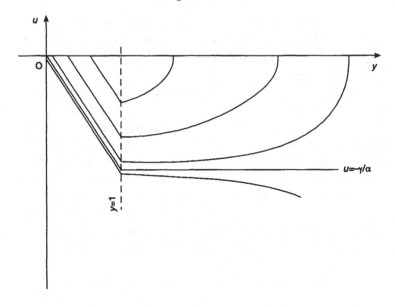

Figure 5.4.1. Part of the direction field of differential equation (5.4.1).

$y < 1$, $\dot{u} = -\alpha$. When $y > 1$, $\dot{u} = 0$ on the line $u = -\gamma/\alpha$, which is itself an integral curve. As before, the y-axis ($u = 0$) is the locus of infinite slope. Each of the integral curves sketched in figure 5.4.1 represents a possible traveling-wave solution.

From this plethora of traveling-wave solutions, one solution stands out. This one we identify by considering how, in the laboratory practice of applied superconductivity, one creates traveling waves in an uncooled superconducting wire. An initial temperature rise $c(z, 0)$ similar to that sketched in figure 5.4.2(a) is established in the wire. If it is large enough, then as time advances, it rises and broadens as shown in figure 5.4.2(a). If we plot $u \equiv c_z$ versus $y \equiv c$ for the right halves of the profiles of figure 5.4.2(a), we obtain figure 5.4.2(b).

If we assume that eventually the initial distribution matures into two traveling waves moving in opposite directions, then the profiles of figure 5.4.2(b) must approach those of figure 5.4.1. Now since $c(0, t) = y_{max}$ keeps increasing, the y–u profiles

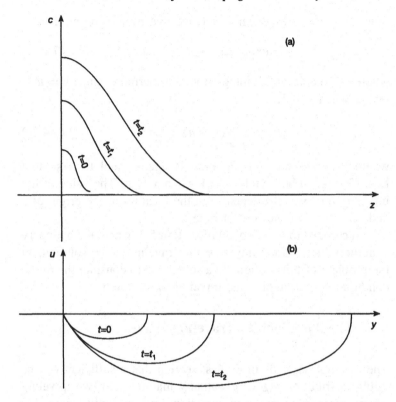

Figure 5.4.2. (a) The temporal development of an initial temperature rise $c(z, 0)$ (right half). (b) The profiles of $u \equiv c_z$ versus $y \equiv c$ for the temperature distributions shown in Fig. 5.4.2(a).

grow continuously until they approach the line $u = -\gamma/\alpha$ for $y > 1$. Furthermore, for the profiles of figure 5.4.2(a), as $z \to \infty$, $y = c \to 0$ and $u = c_z \to 0$. Hence, for $y < 1$, the profiles must ultimately pass through the origin and thus approach the line $u = -\alpha y$. Since the lines $u = -\gamma/\alpha$ and $u = -\alpha y$ must join at $y = 1$, we find at once that

$$\alpha = \gamma^{1/2} \qquad (5.4.2a)$$

Since $\dot{y} = u = -\alpha y$ when $y < 1$, we find by integrating that

$$y = \exp[-\alpha(x - a)] \qquad y < 1 \qquad (5.4.2b)$$

where a is a constant of integration. Furthermore, since $\dot{y} = u = -\gamma/\alpha$ when $y > 1$,

$$y = 1 - \frac{\gamma}{\alpha}(x - a) \qquad y > 1 \qquad (5.4.2c)$$

where the constant of integration is again equal to a so that Eqs. (5.4.2b, c) both yield $y = 1$ when $x = a$. The value of the integration constant a depends on the location of the origin of z and the choice of the zero of time t.

In contrast to the problem of section 5.3, where the boundary conditions determined uniquely the traveling-wave solution, in the problem of this section we also needed to consider the initial condition in determining the traveling-wave solution.

5.5 The Approach to Traveling Waves

The assumption made in the last section that a sufficiently large initial distribution $c(z, 0)$ eventually matures into two traveling waves moving in opposite directions is commonly made but nonetheless deserves to be studied in some detail. We can do this with relative economy of labor by choosing $Q(c) = \gamma c$, in which case the partial differential equation (5.1.2) is analytically solvable on the infinite interval $-\infty < z < \infty$. Furthermore, if we redefine the units of z and t by replacing $\gamma^{1/2}z$ by z and γt by t, we need only deal with the case $\gamma = 1$, which we do in this section and the next.

First, let us determine the traveling-wave solutions. The associated differential equation now takes the form

$$\dot{u} = -\alpha - \frac{y}{u} \qquad (5.5.1)$$

for all y, $0 < y < \infty$. Shown in figure 5.5.1 is the direction field of Eq. (5.5.1). If we restrict our considerations to traveling

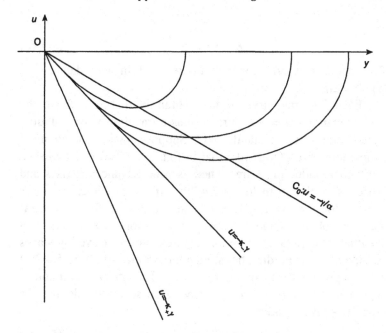

Figure 5.5.1. Part of the direction field of differential equation (5.5.1). C_0 is the locus of zero slope.

waves $y(x)$ that vanish at infinity (where $u = \dot{y}$ must also vanish) then we want solutions passing through the origin O of the (y, u)-plane. In the fourth quadrant, these solutions fall into three subfamilies separated by two separatrices, which have the equations

$$u = -\kappa_\pm y \qquad (5.5.2a)$$

where $\kappa_\pm$ are the roots of the quadratic equation

$$\kappa^2 - \alpha\kappa + 1 = 0 \qquad (5.5.2b)$$

It follows from Eq. (5.5.2b) that $\kappa_+ + \kappa_- = \alpha$ and $\kappa_+\kappa_- = 1$ so that

$$\frac{\kappa_+\kappa_-}{\kappa_+ + \kappa_-} = \frac{1}{\alpha} \qquad (5.5.2c)$$

Thus
$$\kappa_+ > \kappa_- > \frac{1}{\alpha} \tag{5.5.2d}$$
The separatrices are drawn in figure 5.5.1 in accordance with Eq. (5.5.2d).

Each integral curve of the subfamily that lies above the upper separatrix $u = -\kappa_- y$ represents a traveling-wave solution. Again, one solution stands out. Suppose again that an intial temperature rise $c(z, 0)$ like that in figure 5.4.2(a) is established in a superconducting wire. Then as time advances it rises and broadens as shown in figure 5.4.2(a). If we plot $u \equiv cz$ versus $y \equiv c$ for the right halves of the profiles of figure 5.4.2(a) we again obtain figure 5.4.2(b). If, as before, we assume that eventually the inital distribution matures into two traveling waves moving in opposite directions, then the profiles in figure 5.4.2(b) must approach those in figure 5.5.1. Thus they continuously approach the separatrix $u = -\kappa_- y$. This separatrix defines the traveling-wave solution
$$y = \exp[-\kappa_-(x - a)] \tag{5.5.3a}$$
which propagates with speed α related to κ_- by
$$\kappa_- = [\alpha - (\alpha^2 - 4)^{1/2}]/2 \tag{5.5.3b}$$
Note that α must be ≥ 2. If $\alpha < 2$, no separatrices appear in the direction field (5.5.1) and the integral curves wind around the origin O. Then they correspond to oscillatory solutions $y(x)$, which, in applied superconductivity at least, play no role.

Even though we have used the initial condition, we are still left with an infinitude of traveling-wave solutions. Even more subtle considerations are required to determine what value of the propagation velocity α actually occurs in practice.

5.6   The Approach to Traveling Waves (part 2)

When $Q(c) = c$, Eq. (5.1.2) can be written as the ordinary diffusion equation
$$(e^{-t}c)_t = (e^{-t}c)_{zz} \tag{5.6.1}$$

It follows that the Green's function (the solution for a pulsed delta-function source $\delta(t)\,\delta(z)$ at the origin) is

$$c = (4\pi t)^{-1/2} \exp\left(\frac{-z^2}{4t} + t\right) \qquad (5.6.2)$$

The time-dependent solution (5.6.2) is simple enough that we can study analytically how it is related to the traveling-wave solutions of the previous section. The local velocity of propagation v of a fixed value of c is given by

$$v = -\frac{c_t}{c_z} \qquad (5.6.3a)$$

and when c is given by Eq. (5.6.2),

$$v = \frac{2t-1}{z} + \frac{z}{2t} \qquad (5.6.3b)$$

When $t > 1/2$, v has a single minimum when plotted against z. This minimum, which occurs at

$$z_{\min} = [2t(2t-1)]^{1/2} \qquad (5.6.3c)$$

has a value

$$v_{\min} = 2\left(\frac{2t-1}{2t}\right)^{1/2} \qquad (5.6.3d)$$

The breadth Δz of this minimum is of the order $2t$, and the minimum itself corresponds to a value of c given by

$$c_{\min} = \left(\frac{e}{4\pi t}\right)^{1/2} \qquad (5.6.3e)$$

Thus when $t \gg 1$, there is a broad minimum in v for $z \approx 2t$ over which v has values rather close to $v_{\min} \approx 2$.

To make the situation completely transparent, figures 5.6.1a–f give plots of c versus z, v versus z, and v versus c for $t = 5$ and $t = 20$. These plots show that there is a steadily broadening z-region in the expanding c-profile for which the propagation velocity v draws ever closer to the value 2. Over the central

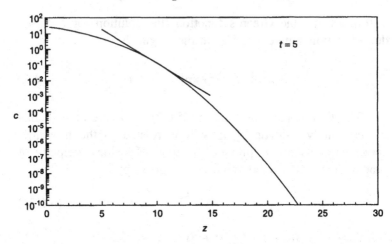

Figure 5.6.1a. A plot of c versus z for $t = 5$. The straight line segment is proportional to e^{-z}.

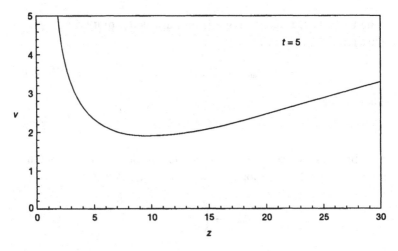

Figure 5.6.1b. A plot of v versus z for $t = 5$.

part of that z-region c is proportional to e^{-z} just as it is in the traveling-wave solution corresponding to $v = 2$. Furthermore, if we track one particular value of c (as is done in experiments in applied superconductivity), it advances with a propagation

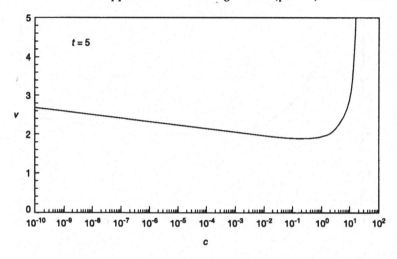

Figure 5.6.1c. A plot of v versus c for $t = 5$.

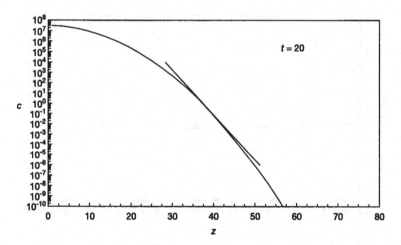

Figure 5.6.1d. A plot of c versus z for $t = 20$. The straight line segment is proportional to e^{-z}.

velocity that asymptotically approaches 2. When $\gamma \neq 1$, this asymptotic limit is $2\gamma^{1/2}$.

The most noteworthy thing about this example is the somewhat unusual way in which the time-dependent solution is

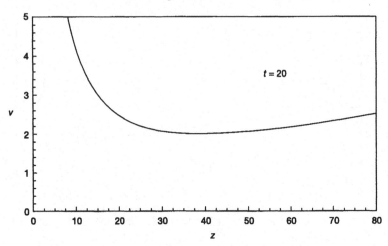

Figure 5.6.1e. A plot of v versus z for $t = 20$.

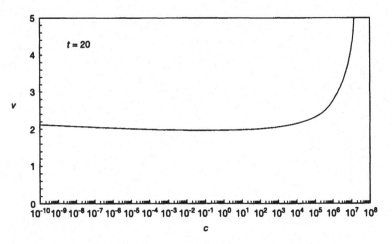

Figure 5.6.1f. A plot of v versus c for $t = 20$.

related to the traveling-wave solution, which seems to develop in the middle of the time-dependent solution and grow outwards.

We can extend the foregoing reasoning to initial conditions $c(z, 0)$ more general than the pulsed delta-function source $\delta(t)\,\delta(z)$ at the origin as follows. Since Eq. (5.1.2) is linear when $Q(c) = c$, we can use the Green's function (5.6.2) to write the solution $c(z, t)$ corresponding to any initial condition $c(z, 0)$ in the integral form

$$c(z, t) = (4\pi t)^{-1/2} e^t \int\limits_{-\infty}^{\infty} \exp\left[-\frac{(z - z')^2}{4t}\right] c(z', 0) \, dz' \quad (5.6.4)$$

If the initial condition $c(z, 0)$ is confined to a finite interval of z, which for convenience we take to be $0 > z > -a$, then we can write for $z > 0$,

$$c(z, t) < (4\pi t)^{-1/2} \exp\left(-\frac{z^2}{4t} + t\right) \int\limits_{-a}^{0} c(z', 0) \, dz' \quad (5.6.5)$$

If we now track one particular value of c, then, by an extension of our earlier argument, it cannot advance with a propagation velocity that asymptotically exceeds $2\gamma^{1/2}$.

It is also possible to extend the reasoning of this section to more general source functions $Q(c)$ in the following way. The solutions of the partial differential equation (5.1.2) obey the following ordering theorem: let $c_1(z, t)$ and $c_2(z, t)$ be two solutions of Eq. (5.1.2) on the interval $a < z < b$ belonging, respectively, to source functions $Q_1(c) > Q_2(c)$. If c_1 and c_2 have the same boundary and initial conditions $c(a, t)$, $c(b, t)$ and $c(z, 0)$, then $c_1(z, t) > c_2(z, t)$. The proof of this theorem follows the procedure outlined in section 4.5; some details are given in note 3 at the end of the chapter. Thus, if the source function $Q(c)$ can be bounded from above by γc and if the initial condition is confined to a finite interval $a < z < b$, then when we track one particular value of c, it cannot advance with a propagation velocity that asymptotically exceeds $2\gamma^{1/2}$. The specific formulas (5.4.2a) and (5.3.4b) (for $\gamma > 2$, which corresponds to an advancing wave front) both conform to this result.

5.7   A Final Example

If $Q(c) = cW(c)$, with $W(c)$ still given by Eq. (5.2.5), it turns out
that there are no traveling-wave solutions that vanish at infinity.
Then, Eq. (5.2.2) becomes

$$\dot{u} = -\alpha \qquad\qquad y < 1 \qquad\qquad (5.7.1a)$$

$$\dot{u} = -\alpha - \frac{y}{u} \qquad y > 1 \qquad\qquad (5.7.1b)$$

Figure 5.7.1 shows the direction field of Eq. (5.7.1) when $\alpha \geq 2$.
When $y > 1$, the separatrices are again

$$u = -\kappa_{\pm} y \qquad\qquad (5.7.2a)$$

where $\kappa_{\pm}$ are the roots of the quadratic equation

$$\kappa^2 - \alpha\kappa + 1 = 0 \qquad\qquad (5.7.2b)$$

Suppose again that an intial temperature rise $c(z, 0)$ like that
in figure 5.4.2a is established in a superconducting wire. By
repeating our earlier reasoning, we see that if a traveling-wave
solution is approached asymptotically, its velocity must be such
that the two solutions $u = -\alpha y$ and $u = -\kappa_- y$ are equal at
$y = 1$. But there is no solution to the equation $\alpha = \kappa_-$. So there
is no traveling-wave solution for which $y(\infty) = 0$ when $\alpha \geq 2$.

When $\alpha < 2$, Eq. (5.7.2b) has no real solutions, and the
separatrices $u = -\kappa_{\pm} y$ disappear in the direction field (5.7.1).
All of the integral curves of Eqs. (5.7.1) in the fourth quadrant
of the (y, u)-plane then eventually intersect the y-axis ($u = 0$)
at some value $y > 1$. (For a proof of this, see note 4.) We see
therefore that there is no traveling-wave solution $y(x)$ for which
$y(\infty) = 0$ and $y(-\infty) = \infty$.

What then happens to our initial temperature distribution
$c(z, 0)$ as time goes on? If we track very large values of c,
it should not matter that $Q(c)$ vanishes when $c < 1$. Thus we
expect large values of c to propagate with the asymptotic velocity
$v = 2$. If smaller values of c propagated with smaller asymptotic
velocities, eventually large gradients c_z would be created that

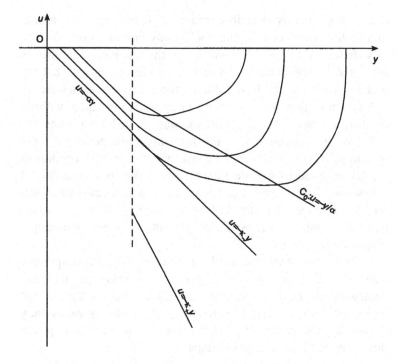

Figure 5.7.1. Part of the direction field of differential equation (5.7.1) when $a > 2$. C_0 is the locus of zero slope.

would tend to push these small values of c to higher velocities. It is my surmise that the situation is not qualitatively different from that described in section 5.6.

5.8 Concluding Remarks

In the problem of section 5.3, stability considerations led us to consider traveling-wave solutions $y(x)$ having specified flat asymptotes. The existence of these asymptotes was enough to determine a unique traveling-wave solution. In a different problem, dealt with in section 5.4, traveling-wave solutions $y(x)$ were identified that do not have a flat asymptote at $x = -\infty$. By

choosing physically plausible initial conditions $c(z, 0)$ confined
to a finite interval of z (figure 5.4.2), only one of these solutions
was shown to be possible. Implicit in the qualitative arguments
presented in both sections 5.3 and 5.4 is the assumption that the
initial distribution $c(z, 0)$ eventually matures (if it does not decay)
into two traveling waves moving in opposite directions without
change of shape. This assumption was studied in sections 5.5
and 5.6. In section 5.5, many traveling-wave solutions were
identified, all of which were consistent with initial conditions
$c(z, 0)$ confined to a finite interval of z. But in section 5.6, it
was shown that the time-dependent solutions connected with such
initial conditions have the following property: if we track one
particular value of c, it asymptotically advances with the unique
propagation velocity $2\gamma^{1/2}$.

There are two practical advantages of traveling-wave
solutions. The first is that, like similarity solutions, they are
comparatively easy to calculate. The second advantage is that
initial distributions $c(z, 0)$ that do not immediately decay may
eventually mature into two traveling waves moving in opposite
directions without change of shape.

Notes

Note 1:   The stability treatment based on Eq. (5.3.1) is too narrow because
it deals only with spatially uniform perturbations $\varepsilon(t)$. If, more generally, we
subsitute $c = \underline{c} + \varepsilon(z, t)$ into Eq. (5.1.2), we obtain, to lowest order in ε,

$$\varepsilon_t = \varepsilon_{zz} + \varepsilon \left(\frac{dQ}{dc} \right)_{c=\underline{c}} \tag{N1.1}$$

If we Fourier transform Eq. (N1.1) with respect to z, i.e. if we set

$$\underline{\varepsilon}(k, t) = \int\limits_{-\infty}^{\infty} \varepsilon(z, t) \exp(-ikz) \, dz \tag{N1.2}$$

we find

$$\underline{\varepsilon}_t = \left[-k^2 + \left(\frac{dQ}{dc} \right)_{c=\underline{c}} \right] \underline{\varepsilon} \tag{N1.3}$$

As before, if $(dQ/dc)_{c=\underline{c}} < 0$, all Fourier components decay and thus all initial perturbations decay. If $(dQ/dc)_{c=\underline{c}} > 0$, then Fourier components for which $k^2 < (dQ/dc)_{c=\underline{c}}$ grow. If $(dQ/dc)_{c=\underline{c}} > 0$, then initial perturbations that contain only frequencies for which $k^2 > (dQ/dc)_{c=\underline{c}}$ decay. But the perturbations caused by random thermal fluctuations can hardly be expected to have such a special character. Thus we continue to conclude that the state $c = \underline{c}$ is stable or unstable according to whether $(dQ/dc)_{c=\underline{c}}$ is < 0 or > 0, respectively.

Note 2: By sketching the direction field of Eq. (5.3.2a) near the origin O, the reader may convince himself that the origin is a saddle-point singularity at which two separatrices intersect at the origin. None of the other integral curves, which are divided into four subfamilies by the separatrices, pass through the origin. By direct substitution, we find that the curves $u = -\kappa_{\pm}y$, where $\kappa_{\pm}$ are the roots of Eq. (5.3.2b), satisfy Eq. (5.3.2a) and pass through the origin. Thus they must be the two separatrices. The one we seek here is that for which $y < 0$, namely, $u = -\kappa_{+}y$. Similar reasoning applies to the solution $u = -\kappa_{-}(y - \gamma)$ of Eq. (5.3.3a).

Note 3: As in section 4.5, let $u = c_1 - c_2$ and subtract Eq. (5.1.2) written for c_2 from Eq. (5.1.2) written for c_1:

$$u_t = u_{zz} + Q_1(c_1) - Q_2(c_2) \tag{N3.1}$$

Now

$$Q_1(c_1) - Q_2(c_2) = Q_1(c_1) - Q_1(c_2) + Q_1(c_2) - Q_2(c_2) \tag{N3.2a}$$

$$= \left(\frac{dQ_1}{dc}\right)u + \delta Q \tag{N3.2b}$$

where $\delta Q \equiv Q_1(c_2) - Q_2(c_2) > 0$ and dQ_1/dc is evaluated at an argument lying between c_1 and c_2. If as before we introduce $v = ue^{-\gamma t}$, we obtain

$$v_t = v_{zz} + \left(\frac{dQ_1}{dc} - \gamma\right)v + \delta Q e^{-\gamma t} \tag{N3.3}$$

Since $\delta Q > 0$, the argument of section 4.5 now goes through without change.

Note 4: Equation (5.7.1b) is invariant to the stretching group $u' = \lambda u$, $y' = \lambda y$. It can therefore be integrated with the help of Lie's integrating factor (section 2.1). When $\alpha < 2$, the result is

$$\psi = \ln[(u^2 + \alpha uy + y^2)^{1/2}] - \alpha(4 - \alpha^2)^{1/2}\arctan\left[\frac{2u + \alpha y}{y(4 - \alpha^2)^{1/2}}\right] \tag{N4.1}$$

where ψ is a constant that labels the various integral curves. When $u = 0$ and $y = y_{max}$,

$$\psi = \ln y_{max} - \alpha(4 - \alpha^2)^{1/2} \arctan\left[\frac{\alpha}{(4 - \alpha^2)^{1/2}}\right] \qquad \text{(N4.2)}$$

so that each integral curve intersects the y-axis as noted earlier.

The integral curve that joins the line $u = -\alpha y$ at $y = 1$ has the value of ψ given by

$$\psi_* = \alpha(4 - \alpha^2)^{1/2} \arctan\left[\frac{\alpha}{(4 - \alpha^2)^{1/2}}\right] \qquad \text{(N4.3)}$$

so that for this integral curve

$$y_{max} = \exp(2\psi_*) \qquad \text{(N4.4)}$$

Thus we have shown by direct calculation that there is no traveling-wave solution $y(x)$ for which $y(\infty) = 0$ and $y(-\infty) = \infty$.

Problems for Chapter 5

5.1 Determine the traveling-wave solutions of the source-free diffusion equation $c_t = c_{zz}$. Use formula (5.6.3a) for the local velocity v of propagation of a fixed value of c to show that $v = \alpha$ for all values of c for a traveling wave solution.

5.2 What condition must V satisfy for Eqs. (4.9.1a, b) to have traveling-wave solutions?

5.3 Determine the traveling-wave solutions of the wave equation $c_{tt} = c_{zz}$. Is there something peculiar about the principal differential equation? Construct infinite-medium ($-\infty < z < \infty$) traveling-wave solutions that correspond to the initial conditions $c(z, 0) = u(z)$, $c_t(z, 0) = 0$. Now construct infinite-medium traveling-wave solutions that correspond to the initial conditions $c(z, 0) = 0$, $c_t(z, 0) = v(z)$. If you add these solutions you get d'Alembert's classical infinite-medium solution of the wave equation for the intial conditions $c(z, 0) = u(z)$, $c_t(z, 0) = v(z)$.

5.4 J. D. Murray [Mu-89] has used the equation $c_t = c(1 - c) + (cc_z)_z$ to model density-dependent population diffusion with logistic population growth. Verify that it is invariant to the family of groups (5.1.1a–c). Find the principal differential equation of the traveling-wave solutions and using the associated group (5.1.8a, b) and Lie's reduction theorem, reduce the order of the principal equation. Murray shows that the singularities of the reduced equation are at O: (0, 0), P: (1, 0) and Q: (0, −α). Murray tells us that for a certain value of α there is a straight-line solution of the reduced equation joining P and Q. Find it and determine the traveling-wave solution to which it corresponds.

5.5 Suppose a partial differential equation is invariant to the family F of groups (5.1.1a–c) as well as to another group G. Derive sufficient conditions on the coefficients of the infinitesimal transformation of G that allow the existence of another group invariance of the principal differential equation of the traveling-wave solutions besides invariance to the associated group (5.1.8a, b).

5.6 Find a group of stretching transformations that leave Eqs. (5.4.1a, b) invariant. (You need to transform the eigenvalue α and the constant γ as well as the variables u and y.) Using this group, show that α is proportional to $\gamma^{1/2}$ (as the result (5.4.2a) of an exact calculation shows). Can you generalize this argument to Eq. (5.2.2) when $Q(y)$ is replaced by $\gamma Q(y)$?

5.7 According to Murray [Mu-89], a model for the motion of a colony of bacteria moving into a food source on a long, thin tape and consuming it as they go is

$$b_t = \left(b_z - \frac{2bn_z}{n}\right)_z \qquad n_t = -b \tag{1}$$

where b is the density of bacteria, n is the density of nutrients, z is the position coordinate and t is the time. This system of equations is invariant to the family of groups

$$b' = b \qquad n' = n \qquad t' = t + \lambda \qquad z' = z + \alpha\lambda \tag{2}$$

Write the ordinary differential equations corresponding to the traveling-wave solution $b = B(x)$, $n = N(x)$, where $x = z - \alpha t$. These ordinary differential equations are invariant to the associated group (5.1.8a, b). Equations (1) are also invariant to the group

$$z' = z \qquad t' = t \qquad n' = \lambda n \qquad b' = \lambda b \qquad (3)$$

Do the infinitesimal operators of groups (2) and (3) commute? What is the action of group (3) on the principal ordinary differential equations of the traveling-wave solution? Find explicit expressions for the traveling-wave solution.

5.8 Sketch the direction field of Eq. (5.2.2) when $Q(c)$ is given by

$$Q(c) = 0 \qquad c < 1$$
$$Q(c) = c - 1 \qquad c > 1$$

How does the asymptotic form of $c(x)$, $x = z - \alpha t$, when $c \gg 1$, depend on α?

6

Approximate Methods

6.1 Introduction

As mentioned previously, the advantage of invariant solutions is that they are easier to calculate than other solutions. In the comparatively simple cases dealt with in this book, the invariant solutions are calculated by solving ordinary rather than partial differential equations.

Invariant solutions seem to have a disadvantage, however, for if there is the slightest variation in the conditions they require, they no longer apply. For example, in the clamped-temperature problem of the illustrative example given in section 4.2, if the pipe is not semi-infinite but, rather, finite and of length L, so that then the boundary and initial conditions are $c(0, t) = 1$, $c(z, 0) = 0$ and $c(L, t) = 0$, then the invariant similarity solution of Eq. (3.3.11) is no longer correct. However, for early times, i.e. for small enough t, the disturbance from the initial condition $c = 0$ caused by suddenly clamping the temperature at the front face ($z = 0$) at the value $c = 1$ is substantial only close to $z = 0$. In an anthropomorphic manner of speaking that is often used to describe this situation, we may say that for small enough t, the partial differential equation does not yet know that the pipe is finite. So we expect that for short enough times, the similarity solution for an infinite pipe will be a good approximation to the solution for a finite pipe. It is, so to speak, an early-time asymptote.

For late times, on the other hand, the solution of the clamped-temperature problem approaches the steady-state solution of the superfluid diffusion equation, namely, $c = 1 - z/L$, which is therefore a late-time asymptote. These two asymptotes then can be used to determine the early- and late-time behavior of any quantities of interest. For example, the heat flux at the front face $[-c_z(0, t)]^{1/3}$ is given by $(3/4t)^{1/4}$ for early times and $(1/L)^{1/3}$ for late times. If we interpolate graphically between these two asymptotes we can obtain an engineering estimate of $[-c_z(0, t)]^{1/3}$ for all t.

The traveling-wave solutions of the previous chapter may also be considered as late-time asymptotes and thus tell us about the behavior of solutions that, on account of their initial conditions, cannot strictly be traveling waves.

If the partial differential equation admits ordering theorems, the invariant solutions may provide bounds on other solutions that produce useful information. We have seen an example of this in section 4.4 concerning the asymptotic behavior of solutions and another example in section 5.6 concerning bounds on the velocity of propagation.

The use of invariant solutions as asymptotes or as bases for comparison greatly extends their practical utility. But there are yet further situations in which invariant solutions may give us valuable information about the true solution. Consider, for example, the clamped-temperature problem for the superfluid diffusion equation $c_t = (c_z^{1/3})_z$ in a semi-infinite pipe in which $c(0, t) = f(t)$, where $f(t)$ is a slowly varying function of time. If $f(t)$ were constant, the solution would be the similarity solution of Eq. (4.2.7). If $f(t)$ is slowly varying, we may surmise that this similarity solution is close to the correct solution. We show in the next section how to use the similarity solution as a jumping-off point for calculating an improved solution.

6.2 Superfluid Diffusion Equation with a Slowly Varying Face Temperature

When the face temperature $c(0, t)$ is clamped, $c(z, t)$ equals $y(x)$ where $x = z/t^{3/4}$ and $y(x)$ satisfies the differential equation

$$\frac{4}{3}\frac{d(\dot{y}^{1/3})}{dx} + x\dot{y} = 0 \qquad (6.2.1)$$

and vanishes at $x = \infty$. When $c(0, t) = f(t)$, a slowly varying function of time, we try a solution of the form

$$c(z, t) = f(t)\, y\left[\frac{z}{p(t)}\right] \qquad (6.2.2)$$

where $y(x)$ is the function determined by Eq. (6.2.1) and $p(t)$ is a function yet to be determined.

The trial solution (6.2.2) obeys the boundary and intial conditions $c(0, t) = f(t)$, $c(z, 0) = 0$ and $c(\infty, t) = 0$, but it does not, indeed it cannot, obey exactly the superfluid diffusion equation

$$c_t = (c_z^{1/3})_z \qquad (6.2.3)$$

But if we substitute the trial solution (6.2.2) into Eq. (6.2.3) and drop terms of the order of df/dt compared with terms of order f (since f is a slowly varying function of time), we obtain the equation

$$\frac{d(p^{4/3})}{dt} = f^{-2/3} \qquad (6.2.4a)$$

from which we can determine $p(t)$ in terms of $f(t)$. As we shall see subsequently, however, this direct substitution procedure is less accurate than the following procedure, in which we integrate Eq. (6.2.3) over z from 0 to ∞ to obtain an integral relation sufficient to determine $p(t)$ in terms of $f(t)$, namely

$$\frac{d}{dt}\int_0^\infty c\, dz = [-c_z(0, t)]^{1/3} \qquad (6.2.4b)$$

When we substitute Eq. (6.2.2) into Eq. (6.2.4b) we find

$$\frac{d}{dt}\left[f(t)\, p(t) \int_0^\infty y(x)\, dx \right] = \left[-\frac{\dot{y}(0)\, f(t)}{p(t)} \right]^{1/3} \qquad (6.2.5)$$

Now it follows from integrating Eq. (6.2.1) over x from 0 to ∞ and then integrating once by parts that

$$\int_0^\infty y(x)\, dx = \tfrac{4}{3}[-\dot{y}(0)]^{1/3} \qquad (6.2.6)$$

so that Eq. (6.2.5) becomes

$$\frac{d}{dt}[f(t)p(t)] = \frac{3}{4}\left[\frac{f(t)}{p(t)} \right]^{1/3} \qquad (6.2.7a)$$

or after a slight rearrangement

$$(fp)^{\frac{1}{3}}\frac{d(fp)}{dt} = \frac{3f^{2/3}}{4} \qquad (6.2.7b)$$

Equation (6.2.7b) can be integrated subject to the boundary condition $p(0) = 0$ (since for small t we expect $p \approx t^{3/4}$) to give

$$p(t) = \frac{\left\{ \int_0^t [f(t')]^{2/3}\, dt' \right\}^{3/4}}{f(t)} \qquad (6.2.8)$$

Using this value of $p(t)$ we then find from Eq. (6.2.2) that the heat flux at the front face is

$$[-c_z(0, t)]^{1/3} = [-\dot{y}(0)]^{1/3}[f(t)]^{2/3}\left\{ \int_0^t [f(t')]^{2/3}\, dt' \right\}^{-1/4}$$
$$(6.2.9)$$

It follows from the solution (3.3.11) that when $y(0) = 1$, $[-\dot{y}(0)]^{1/3} = (3/4)^{1/4}$.

We can check the accuracy of formula (6.2.9) by comparing it with the results in problems we can solve exactly. These latter problems are precisely the similarity solutions themselves. Suppose therefore that $f(t) = t^{\alpha/\beta}$. Then according to Eqs. (6.2.2) and (6.2.9)

$$c(0, t) = t^{\alpha/\beta} y(0) \qquad (6.2.10a)$$

$$[-c_z(0, t)]^{1/3} = t^{(\alpha-1)/3\beta} [-\dot{y}(0)]^{1/3} \left(1 + \frac{2\alpha}{3\beta}\right)^{1/4} \quad (6.2.10b)$$

it should be remembered that here although α and β can have any values consistent with the linear constraint (4.1.2), $y(x)$ is the similarity solution belonging to the clamped-temperature problem (Eq. (3.3.11)). Thus

$$\frac{[-c_z(0, t)]^{1/3}}{[c(0, t)]^{1/2}} = \left(\frac{3\beta + 2\alpha}{3\beta}\right)^{1/4} \frac{[-\dot{y}(0)]^{1/3}}{[y(0)]^{1/2}} t^{-1/4} \quad (6.2.10c)$$

Owing to the invariance of the principal differential equation for $y(x)$ to the associated group (4.2.10), the ratio $[-\dot{y}(0)]^{1/3}/[y(0)]^{1/2}$ is independent of $y(0)$ and is the same for all similarity solutions belonging to a particular choice of family parameters α and β. When $\alpha = 0$ and $\beta = 4/3$ (clamped-temperature problem), this ratio is $(3/4)^{1/4}$ (cf. Eq. (3.3.11)). Thus, finally, our approximate formula (6.2.10c) becomes

$$\frac{[-c_z(0, t)]^{1/3}}{[c(0, t)]^{1/2}} = \left(\frac{3\beta + 2\alpha}{4\beta}\right)^{1/4} t^{-1/4} \quad (6.2.11)$$

The exact result for the same problem is

$$\frac{[-c_z(0, t)]^{1/3}}{[c(0, t)]^{1/2}} = \frac{[-\dot{y}(0)]^{1/3}}{[y(0)]^{1/2}} t^{-1/4} \quad (6.2.12)$$

where now $y(x)$ is the similarity solution belonging to the family parameters α and β. Thus the approximate form (6.2.11) has the right functional dependence and differs at most by the value of the constant. The direct substitution procedure mentioned at the beginning of this section leads to a similar formula, namely

$$\frac{[-c_z(0, t)]^{1/3}}{[c(0, t)]^{1/2}} = \left(\frac{3\beta - 2\alpha}{4\beta}\right)^{1/4} t^{-1/4} \quad (6.2.13)$$

Table 6.1. The exact coefficient $[-\dot{y}(0)]^{1/3}/[y(0)]^{1/2}$ and the approximate values given by Eqs. (6.2.11) and (6.2.13).

α	β	Exact	Equation (6.2.11)	Equation (6.2.13)
0	4/3	0.9306	0.9306	0.9306
1	2	1.0958	1.0000	0.8409
2	8/3	1.1614	1.0299	0.7825
4	4	1.2189	1.0574	0.7071
10	8	1.2700	1.0829	0.5946

In another work [Dr-90], I have calculated, by numerical integration of the associated differential equation, the exact coefficients $[-\dot{y}(0)]^{1/3}/[y(0)]^{1/2}$ of $t^{-1/4}$ for several problems with various values of α and β. These results, together with the approximate values from Eqs. (6.2.11) and (6.2.13), are shown in table 6.1. As noted at the beginning of this section, the integral method is more accurate than the direct substitution method.

As a final word in this section, let me point out that similarity solutions other than that for the clamped-temperature problem can be used as the jumping-off point for calculating an improved solution. The one chosen should, of course, be the one that most closely matches the conditions of the problem.

6.3 Ordinary Diffusion with a Non-Constant Diffusion Coefficient

The problem of the last section involved an invariant partial differential equation but a non-invariant boundary condition. The technique of solution explained there can also be applied to non-invariant partial differential equations that are 'close' to invariant partial differential equations. Consider, for example, the diffusion equation with a temperature-dependent diffusion constant:

$$c_t = (D(c)c_z)_z \tag{6.3.1}$$

We shall be interested in solutions that obey the partial boundary and initial conditions $c(\infty, t) = 0$ and $c(z, 0) = 0$.

When $D = 1$, the similarity solutions of Eq. (6.3.1) are given by

$$c = t^{\alpha/2}y(x) \tag{6.3.2a}$$

where $x = z/t^{1/2}$ and $y(x)$ is a solution of the principal differential equation

$$2\ddot{y} + x\dot{y} - \alpha y = 0 \tag{6.3.2b}$$

The solutions $y(x)$ that vanish at infinity obey the following integral relation obtained by integrating Eq. (6.3.2b) over x from 0 to ∞ and then integrating once by parts:

$$-2\dot{y}(0) = (\alpha + 1)\int_0^\infty y\, dx \tag{6.3.3}$$

If we now integrate Eq. (6.3.1) over z from 0 to ∞, we find

$$\frac{d}{dt}\left(\int_0^\infty c\, dz\right) = -D_0(t)\, c_z(0, t) \tag{6.3.4}$$

where $D_0(t) \equiv D[c(0, t)]$. Let us now choose a trial solution of the form

$$c(z, t) = q(t)y\left(\frac{z}{p(t)}\right) \tag{6.3.5}$$

Then we find

$$c(0, t) = q(t)\, y(0) \tag{6.3.6a}$$

$$c_z(0, t) = \frac{q(t)}{p(t)}\dot{y}(0) \tag{6.3.6b}$$

and

$$\frac{d}{dt}(qp) = (\alpha + 1)D_0(t)\frac{q}{2p} \tag{6.3.6c}$$

Equation (6.3.6c) can be integrated to give

$$p(t) = (\alpha + 1)^{1/2} \frac{\left[\int_0^t D_0 q^2 \, dt'\right]^{1/2}}{q(t)} \qquad (6.3.7)$$

since $p(0) = 0$ (we expect $p(t) \approx t^{1/2}$ for small t). According to Eqs. (6.3.6a, b) the ratio of the heat flux to the temperature rise at the front face is

$$-\frac{D_0(t) \, c_z(0, t)}{c(0, t)} = \left[\frac{D_0(t)}{p(t)}\right]\left[-\frac{\dot{y}(0)}{y(0)}\right] \qquad (6.3.8)$$

The ratio $-\dot{y}(0)/y(0)$ is independent of $y(0)$ since the principal differential equation is linear and thus the ratio depends only on the family parameter α of the base similarity solution $y(x)$.

6.4 Check on the Accuracy of the Approximate Formula (6.3.8)

We can check the accuracy of the approximate formula (6.3.8) by finding a solvable problem to compare it with. One such problem is the clamped-temperature problem $(c(0, t) = 1)$ when $D(c) = e^c$. Now when $\alpha = 0$, the base solution is $c(z, t) = y(x) = \text{erfc}(x/2)$. Thus according to Eq. (6.3.7), since $q(t) = 1$, $p(t) = (et)^{1/2}$ since $D_0 = e$. Since $-\dot{y}(0)/y(0) = \pi^{-1/2}$, the right-hand side of the approximate formula Eq. (6.3.8) is $(e/\pi)^{1/2} t^{-1/2} = 0.930\,19 t^{-1/2}$.

The partial differential equation

$$c_t = (e^c c_z)_z \qquad (6.4.1)$$

is invariant to the family of mixed stretching-translation groups

$$c' = c + \alpha \ln \lambda \qquad (6.4.2a)$$

$$t' = \lambda^\beta t \qquad (6.4.2b)$$

$$z' = \lambda z \qquad (6.4.2c)$$

where

$$\alpha + \beta = 2 \tag{6.4.2d}$$

The invariants of any group of the family are

$$x = \frac{z}{t^{1/\beta}} \tag{6.4.3a}$$

$$y = c - \frac{\alpha}{\beta} \ln t \tag{6.4.3b}$$

The principal differential equation is

$$\beta \frac{d(e^y \dot{y})}{dx} + x\dot{y} - \alpha = 0 \tag{6.4.4a}$$

The clamped-flux problem $c(0, t) = 1$ is characterized by the family parameters $\alpha = 0$, $\beta = 2$ so that for this problem Eq. (6.4.4a) can be written

$$2\frac{d(e^y \dot{y})}{dx} + x\dot{y} = 0 \qquad \text{where} \qquad x = \frac{z}{t^{1/2}} \tag{6.4.4b}$$

Equation (6.4.4b) must be solved with the boundary conditions $y(0) = 1$ and $y(\infty) = 0$.

If $x \gg 1$, then $y \ll 1$, and the factor e^y in the first term of Eq. (6.4.4b) can be neglected. It follows at once that

$$y \sim C \, \mathrm{erfc}(x/2) \qquad \text{for} \qquad x \gg 1 \tag{6.4.5}$$

where C is a constant of integration. To determine C we must undertake a backwards numerical integration of Eq. (6.4.4b) using Eq. (6.4.5) to calculate starting values of y and $\dot{y}$ at some large value of x. The quantity C must be chosen so that $y(0) = 1$ and some trial and error is unavoidable because the associated group is of no help here. A short calculation similar to that in section 4.3 shows that the associated group is

$$y' = y + 2 \ln \lambda \tag{6.4.6a}$$

$$x' = \lambda x \tag{6.4.6b}$$

and we cannot use this group to rescale the value of C if our intial guess is wrong. Four iterations yield $C = 1.581\,78$ and

$\dot{y}(0) = -0.289\,00$ when $y(0) = 1$ and $y(\infty) = 0$. Then

$$-\frac{D_0(t)\,c_z(0, t)}{c(0, t)} = e\frac{-\dot{y}(0)}{y(0)}t^{-1/2} = 0.785\,58\,t^{-1/2} \qquad (6.4.7)$$

since $D_0(t) = e$. The approximate answer $0.930\,19\,t^{-1/2}$ is thus 18% higher than the exact answer (6.4.7).

Problems for Chapter 6

6.1 Derive an approximate formula for $c_z(0, t)$ when $c(z, t)$ satisfies $cc_t = c_{zz}$ and $c(0, t) = F(t)$, $t > 0$; $c(\infty, t) = 0$; $c(z, 0) = 0, 0 < z < \infty$, where F is a slowly varying function of t.

6.2 For the partial differential equation of problem 6.1, calculate an approximate formula for $c(0, t)$ if the boundary and intial conditions are $c_z(0, t) = -G(t)$, $t > 0$; $c(\infty, t) = 0$; $c(z, 0) = 0$, $0 < z < \infty$ and G is a slowly varying function of t.

6.3 A restricted form of the ordinary diffusion equation in *cylindrical* coordinates is $c_t = (rc_r)_r/r$. This partial differential is invariant to the family of groups

$$c' = \lambda^\alpha c \qquad t' = \lambda^2 t \qquad r' = \lambda r$$

The solution corresponding to the boundary and initial conditions $c_r(R, t) = -q$, $t > 0$; $c(\infty, t) = 0$; $c(z, 0) = 0$, $R < r < \infty$ is not invariant to the transformations of the family because the source position R changes.

(a) If we add to the family the transformations

$$q' = \lambda^{\alpha-1}q \qquad R' = \lambda R$$

we embed the problem in a class of problems all of the same kind. For all of these problems $C \equiv c(R, t)$ is a function of only q, R and t, i.e. $C = F(q, R, t)$. Determine the most general form of the function F.

(b) The boundary condition $c_r(R, t) = -q$ is equivalent to the condition $\int_R^\infty rc(r, t)\, dr = Rqt$ (prove it!). When $t \gg 1$ and the c-distribution has spread far from $r = R$, the value of the integral should be little affected if the lower limit is set equal to zero. With the new boundary condition $\int_0^\infty r\, c(r, t)\, dr = Rqt$ replacing the boundary condition $c_r(R, t) = -q$, the solution becomes a similarity solution. Calculate it.

Epilog

Now that the reader has studied the text carefully, it is perhaps worth summing up what has been presented and saying a word about what has not.

The presentation in this book revolves around three central themes: Lie's integrating factor for first-order ordinary differential equations, Lie's reduction theorem for higher-order ordinary differential equations, and the calculation of invariant solutions of partial differential equations. The great advantage of Lie's methods is that they do not depend on the differential equation's being linear. Consequently, I consider Lie's theory to provide the only widely applicable, systematic treatment available for nonlinear differential equations.

One question of prime importance that has been touched on at various points in the text is the difficult question of how to find groups which leave a given differential equation invariant. For a differential equation with an uncomplicated structure, it is often possible to find, more or less at a glance, a simple group to which the equation is invariant. Lie's method of tabulation, mentioned in section 2.6, is a way of making a dictionary of model ordinary differential equations invariant to groups with various assumed coefficient functions. Sections 3.9 and 3.10 present a method based on Noether's theorem of creating a collection of second-order differential equations for which an explicit first integral is known.

Inspection and tabulation have a hit-and-miss character. The method of determining equations touched on in section 3.11 is a direct method for finding groups to which a given differential equation is invariant. Sometimes the method of determining equations can be streamlined (see problem 2.2). It is perhaps worth noting here that the determining equations do not always have a solution: Cohen [Co-11] gives an example of a second-order ordinary differential equation that is not invariant to any group!

Some of the simpler group invariances are connected with the physical symmetry of the underlying problems. For example, when there is no preferred orientation of the coordinate axes, the differential equation is invariant to the rotation group. When there is no preferred location for the origin of a coordinate, the differential equation is invariant to the group of translations of that coordinate. And when there is no preferred scale for a variable (for example, no preferred length in a diffusion problem in a semi-infinite or infinite medium), the differential equation is invariant to the group of stretchings of that variable. Because these physical symmetries are present in wide variety of interesting technological problems, the rotational, translational and stretching groups occur quite often.

When partial differential equations are involved, a similarity solution, such as those discussed in chapters 4 and 5, is often the solution to the problem we face. When it is not, it is sometimes possible to reformulate the physical problem in such a way that its solution becomes a similarity solution. An example of this occurs in the application of the high-temperature superconductors: certain problems can be made invariant to stretching groups by assuming the voltage–current curve of the superconductor to be a power law. While not exact, this assumption is close enough to be useful, and the resulting similarity solution provides valuable information with a minimum investment of calculational labor.

As mentioned in section 6.1, similarity solutions often represent the short-time behavior of more complicated problems. The entrance-region solutions of fluid flow and heat and mass transfer in pipes are typical short-time asymptotes. Sometimes, an acceptable solution may be won by combining short-time and long-time asymptotic solutions graphically. Also mentioned in section 6.1 is the use of similarity solutions to bound the solution of a more complicated problem. The rest of chapter 6 is devoted to the use of similarity solutions as the jumping-off point for improved solutions.

A lot of attention has been paid in chapter 3 to the reduction of second-order ordinary differential equations to first order using Lie's reduction theorem and the subsequent study of the first-order equation by means of its direction field. Even when the solution ultimately must be found by numerical integration, such a study is an invaluable prelude that often shrinks the required calculational effort. Because many of the differential equations of classical physics are second order, this situation is a common one. The same may be said of the similarity solutions treated in chapter 4, where invariance of the partial differential equation to a family of stretching groups makes the principal ordinary differential equation invariant to the associated group and thus reducible to a first-order equation.

The groups considered in this book are groups of point transformations (see Eqs. (1.1.1*a*, *b*)). Lie himself considered more general kinds of transformations, called contact transformations. The reader interested in pursuing the Lie theory of differential equations further may find a simple, direct treatment of this subject in chapter VII of Cohen's book [Co-11]. More extensive, modern treatments of contact transformations (as well as the entire Lie theory of differential equations) can be found in the books of Bluman and Kumei [Bl-89] and Olver [Ol-86]. Recently, even more general kinds of transformations, non-local transformations, have been the subject of study (see, for example, reference [Go-95]). Although these matters are both practically useful and theoretically important, they are, regrettably, beyond the scope of this introductory book.

Appendix A

Linear, First-Order Partial Differential Equations

A.1 Introduction

In this book, we are interested in linear, first-order partial differential equations in one dependent variable z and two independent variables x and y. The general form of such equations is†

$$P(x, y, z) z_x + Q(x, y, z) z_y = R(x, y, z) \qquad (A.1.1)$$

A solution is a function $z(x, y)$ that satisfies Eq. (A.1.1) identically. Such a solution may be given implicitly in the form $f(x, y, z) = c$, where c is a constant. Since $z_x = -f_x/f_z$ and $z_y = -f_y/f_z$, Eq. (A.1.1) is equivalent to the more symmetric equation

$$P(x, y, z) f_x + Q(x, y, z) f_y + R(x, y, z) f_z = 0 \qquad (A.1.2)$$

† Formally speaking, Eq. (A.1.1) is quasilinear because although it is linear in the partial derivatives z_x and z_y, the variable z appears in the functions P, Q and R. Only if z did not appear in the functions P, Q and R would Eq. (A.1.1) be called linear. The theory developed in this appendix applies to quasilinear equations, a class which includes the strictly linear equations.

A.2   Characteristic Curves

If the points O: (x, y, z) and O′: $(x + dx, y + dy, z + dz)$ both lie in the solution surface S: $f(x, y, z) = c$, then to lowest order

$$f_x \, dx + f_y \, dy + f_z \, dz = 0 \qquad (A.2.1)$$

Since this relation holds for all infinitesimal vectors (dx, dy, dz) tangent to the surface S at point O, the vector (f_x, f_y, f_z) must be normal to S at O. But then, according to Eq. (A.1.2), a vector whose x-, y- and z-components are proportional to P, Q and R, respectively, lies in the solution surface S. Thus the *characteristic equations*

$$\frac{dx}{P} = \frac{dy}{Q} = \frac{dz}{R} \qquad (A.2.2)$$

trace out curves that all lie in the solution surface S. These curves are called *characteristic curves*.

The equations (A.2.2) define a characteristic curve through every point (x, y, z) in space. These curves form a two-parameter family (we could label them, for example, according to their intersections with some surface, say, the (x, y)-plane). Any continuous, one-parameter family of such curves defines a surface S′: $g(x, y, z) = c$ and, as we prove next, such a surface is a solution surface of the partial differential equation. For, at every point of such a surface the direction (dx, dy, dz) of the characteristic curve through that point lies in the surface. In that direction,

$$g_x \, dx + g_y \, dy + g_z \, dz = 0 \qquad (A.2.3)$$

But since the increments (dx, dy, dz) in a characteristic direction are given by Eq. (A.2.2), Eq. (A.2.3) becomes

$$Pg_x + Qg_y + Rg_z = 0 \qquad (A.2.4)$$

so that the surface S′: $g(x, y, z) = c$ is a solution surface of the partial differential equation.

Thus, solution surfaces are ruled with characteristic curves and conversely surfaces ruled with characteristic curves are solution surfaces.

A.3   Integrals of the Characteristic Equations

An *integral* of the characteristic equations is a function $u(x, y, z)$ that is constant along a characteristic curve. Practically speaking, such functions are found by integrating Eqs. (A.2.2). For example, if $P = x$, $Q = 1$, and $R = z$, $u = x/z$, $v = xe^{-y}$ and $w = ze^{-y}$ are all integrals of the characteristic equations; but only two of them are independent, for $v = uw$. In general, Eqs. (A.2.2) have only two independent integrals; call them $u(x, y, z)$ and $v(x, y, z)$.

Every characteristic curve is thus labeled by a pair of values (a, b) that are the respective values of u and v on it. A one-parameter family of such curves is then given by any functional relationship $F(a, b) = 0$ between a and b. However, such a one-parameter family of characteristic curves forms a solution surface and vice versa. Thus, the solutions of the partial differential equation are given by

$$F[u(x, y, z), v(x, y, z)] = 0 \qquad (A.3.1a)$$

where F is an *arbitrary function*. By solving Eq. (A.3.1a) for v, we may write

$$v(x, y, z) = G[u(x, y, z)] \qquad (A.3.1b)$$

where G is another arbitrary function.

Appendix B

Riemann's Method of Characteristics

B.1 Introduction

Riemann invented the method of characteristics in the middle of the last century during his study of compressible gas flow. The method is generally applicable to problems involving the propagation of disturbances with finite velocity. Such problems are qualitatively different from diffusion problems, in which a disturbance at one point instantaneously produces some effect everywhere. In propagation problems, space is divided into different regions some of which the propagating disturbance has already reached and some of which are still undisturbed. These regions are separated by moving interfaces on one side of which the quantities of interest (e.g. pressure, density, flow velocity) have been disturbed from their initial values and on the other side of which the quantities of interest still have their undisturbed values.

To show the essential features of Riemann's method, an example will be worked below, namely, the motion of water in a long, narrow, open channel. This introductory description suffices for this book. Readers interested in further study of Riemann's method may consult reference [Co-48], which I recommend highly.

B.2   The Motion of Water in a Long, Narrow, Open Channel

The motion of water in a long, narrow, open channel is described by the coupled partial differential equations

$$h_t + (vh)_z = 0 \tag{B.2.1a}$$

$$v_t + vv_z + h_z = 0 \tag{B.2.1b}$$

where z is the longitudinal position coordinate, t is the time, h is the height of the surface of the water above the channel floor, and v is the longitudinal flow velocity (assumed independent of depth). These equations have been written for simplicity in a system of units in which the water density and the acceleration of gravity equal 1.

Riemann starts by multiplying the first of these equations by an as yet undetermined function κ and adding it to the other equation:

$$v_t + (\kappa h + v)v_z + \kappa h_t + (\kappa v + 1)h_z = 0 \tag{B.2.2}$$

Now, for any function $f(z, t)$ the linear combination $af_t + bf_z$ is proportional to the directional derivative of f along a curve in the (z, t)-plane whose local slope is $dz/dt = b/a$. For by the chain rule, along the curve $z(t)$

$$\frac{df}{dt} = f_t + f_z\left(\frac{dz}{dt}\right) = f_t + f_z\left(\frac{b}{a}\right) = \frac{af_t + bf_z}{a} \tag{B.2.3}$$

Riemann's idea is to choose κ so that the directional derivatives of v and h in Eq. (B.2.2) are in the same direction. This requires

$$\kappa h + v = \frac{\kappa v + 1}{\kappa} \tag{B.2.4}$$

which is satisfied only if $\kappa^2 h = 1$ or $\kappa = \pm h^{-1/2}$. With these values of κ, the only possible common directions of differentiation of both v and h are given by $dz/dt = v \pm h^{1/2}$. Combining the directional derivatives of v and h we find

$$(v \pm 2h^{1/2})_t + (v \pm h^{1/2})(v \pm 2h^{1/2})_z = 0 \tag{B.2.5}$$

Equation (B.2.5) says that the directional derivatives of the quantities $v \pm 2h^{1/2}$ in the respective directions given by $dz/dt = v \pm h^{1/2}$ are zero. Thus the quantities $v \pm 2h^{1/2}$ are conserved (i.e. are constant) along curves with respective slopes $dz/dt = v \pm h^{1/2}$. The conserved quantities $v \pm 2h^{1/2}$ are called *Riemann invariants* and the curves given by $dz/dt = v \pm h^{1/2}$ are called *characteristics*. The curves with $dz/dt = v + h^{1/2}$ are called the *positive* characteristics and the curves with $dz/dt = v - h^{1/2}$ are called the *negative* characteristics. Note carefully that the slope of the positive characteristic is not necessarily positive nor is that of the negative characteristic necessarily negative; the words positive and negative refer to the sign between v and $h^{1/2}$.

The solution of certain problems is facilitated by Riemann's formulation (B.2.5) of the partial differential equations (B.2.1). Consider, for example, the problem obeying the initial conditions $v(z, 0) = 0$, $-\infty < z < \infty$; $h(z, 0) = h_o$, $z > 0$; $h(z, 0) = 0$, $z < 0$. These initial conditions describe a channel with a dam at $z = 0$ filled on one side with still water of depth h_o and empty on the other side. At time $t = 0$, the restraining dam ruptures. The partial differential equations (B.2.1) determine the subsequent motion of the water.

To see how the solution develops, we plot the characteristics in the (z, t)-plane (see figure B.2.1). Such a plot, in which z is the abscissa and t the ordinate, is called a *wave diagram*. Since the disturbance starts at $z = 0$ and $t = 0$ and propagates both upstream and downstream it can be represented by a *fan* of positive† characteristics separating two regions of uniform properties. Region R_1 represents water behind the dam not yet reached by the disturbance, and region R_2 represents a region of space not yet reached by water released by the breaking dam.

The positive characteristics in the fan are straight lines, a result not generally true but true in this problem, as we now show. The negative characteristic through any point in the fan if prolonged in the direction of decreasing time eventually intrudes

---

† If the fan had been chosen to consist of negative characteristics, then the method outlined below would yield $v = -2h^{1/2} < 0$ everywhere in the fan, which does not fit the physical conditions of the problem.

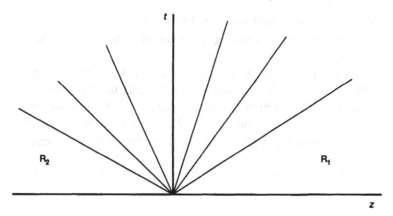

Figure B.2.1. A wave diagram showing a fan of positive characteristics radiating from the origin.

into the region R_1‡. Therefore, at any point in the fan

$$v - 2h^{1/2} = -2h_o^{1/2} \qquad (B.2.6)$$

Since $v + 2h^{1/2}$ is constant on any positive characteristic, we see that both v and h are constant on any positive characteristic in the fan. Thus the direction $(dz/dt)_+ = v + h^{1/2}$ of the characteristic is also constant.

On a positive characteristic radiating from the origin we can now write

$$\frac{z}{t} = v + h^{1/2} \qquad (B.2.7)$$

since the positive characteristics are straight lines. Solving Eqs. (B.2.6) and (B.2.7) for v and h, we find

$$v = \frac{2}{3}\left(\frac{z}{t} - h_o^{1/2}\right) \qquad (B.2.8a)$$

$$h = \frac{1}{9}\left(\frac{z}{t} + 2h_o^{1/2}\right)^2 \qquad (B.2.8b)$$

‡ Note that the slope of the positive characteristic $(dz/dt)_+ = v + h^{1/2}$ is greater than the slope of the negative characteristic $(dz/dt)_- = v - h^{1/2}$.

When $z/t = h_o^{1/2}$, we have $v = 0$ and $h = h_o$, so that $z/t = h_o^{1/2}$ is the equation of the limiting characteristic that separates the fan from region R_1.

The solution (B.2.8) cannot be extended into the region R_2 continuously in both v and h; for beyond the limiting locus $z/t = -2h_o^{1/2}$ on which $h = 0$, h would begin to increase again from zero. On this locus $v = -2h_o^{1/2}$, so that the disturbance moves downstream with a velocity equal to the flow velocity.

Appendix C

The Calculus of Variations and the Euler–Lagrange Equation

C.1 Introduction

The problem that originated the calculus of variations is the so-called *brachistochrone* problem. A frictionless bead slides down a curved wire joining the fixed points A: (a, y_a) and B: (b, y_b) (see figure C.1.1). What shape must the wire take in order that the time of transit be a minimum?

This problem is a generalization of the usual minimax problems of differential calculus because the quantity sought is a function $y(x)$ rather than a single value of x. But, as we shall see, the problem is treated in a way analogous to the way minimax problems are treated in the ordinary differential calculus.

If g is the acceleration of gravity, the velocity of the bead at any point P: (x, y) is given by $v^2 = 2g(y_a - y)$. The increment of time dt it takes the bead to traverse the interval dx at x is $dt = ds/v = [(ds/dx)/v] dx = (1 + \dot{y}^2)^{1/2} dx/v$, where s represents the arc length along the curve $y(x)$. The total time of descent T is then given by

$$T = (2g)^{-1/2} \int_a^b \left(\frac{1 + \dot{y}^2}{y_a - y}\right)^{1/2} dx \qquad \text{(C.1.1)}$$

It is this quantity that we must minimize by correctly choosing a curve $y(x)$ that obeys the conditions $y(a) = y_a$ and $y(b) = y_b$.

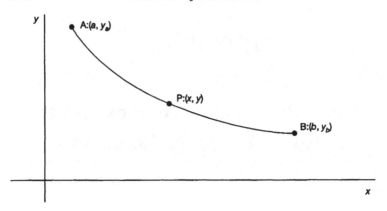

Figure C.1.1. A sketch to explain the brachistochrone problem.

C.2 The Euler–Lagrange Equation

Instead of continuing now with the solution of the brachistochrone problem, we turn instead to a more general treatment
of a whole class of problems that includes the brachistochrone
and is associated with the names of Euler and Lagrange. The
quantity in Eq. (C.1.1) is of the form

$$A[y] = \int_a^b L(x, y, \dot{y}) \, dx \qquad (C.2.1)$$

where the notation on the left-hand side using square brackets
signifies that $A[y]$ is a *functional*, that is, a quantity whose
value is a number but whose argument is a function $y(x)$. The
function L is known as the *Lagrangian*. We seek a function
$y(x)$ for which $y(a) = y_a$ and $y(b) = y_b$ and that makes $A[y]$
an extremum.

We proceed by first converting the problem to one of
ordinary calculus as follows. If $Y(x)$ is the sought-for solution,
Euler and Lagrange consider the family of *trial functions* $y(x) =$
$Y(x) + \varepsilon u(x)$, where ε is a numerical parameter and u is an

arbitrary function except that $u(a) = u(b) = 0$. Then Eq. (C.2.1) becomes

$$A[y] = \int_a^b L(x, Y + \varepsilon u, \dot{Y} + \varepsilon \dot{u}) \, dx \qquad \text{(C.2.2)}$$

so that now A depends only on ε and may be made an extremum by the ordinary method of differential calculus. If we differentiate Eq. (C.2.2) with respect to ε and set $\varepsilon = 0$, $dA/d\varepsilon$ should be zero since $\varepsilon = 0$ is an extremum of A. Thus

$$0 = \frac{dA}{d\varepsilon} = \int_a^b (u L_y + \dot{u} L_{\dot{y}}) \, dx \qquad \text{(C.2.3)}$$

If we integrate the second term in Eq. (C.2.3) by parts we obtain

$$0 = \int_a^b u \left(L_y - \frac{dL_{\dot{y}}}{dx} \right) dx + u L_y \Big|_a^b \qquad \text{(C.2.4)}$$

The integrated term on the right-hand side vanishes because $u(a) = u(b) = 0$. Now if $L_y - dL_{\dot{y}}/dx$ is not zero for all $x, a < x < b$, then by choosing $u(x)$ to have the same sign as $L_y - dL_{\dot{y}}/dx$ we can achieve a positive integral and thus a contradiction. Therefore,

$$\boxed{\frac{dL_{\dot{y}}}{dx} - L_y = 0} \qquad \text{(C.2.5)}$$

which is the celebrated Euler–Lagrange equation for the extremizing function $Y(x)$.

C.3   The Brachistochrone Concluded

In the case of the brachistochrone problem,

$$L = (1 + \dot{y}^2)^{1/2} (y_a - y)^{-1/2} \qquad \text{(C.3.1a)}$$
$$L_{\dot{y}} = \dot{y}(1 + \dot{y}^2)^{-1/2} (y_a - y)^{-1/2} \qquad \text{(C.3.1b)}$$
$$L_y = \tfrac{1}{2}(1 + \dot{y}^2)^{1/2} (y_a - y)^{-3/2} \qquad \text{(C.3.1c)}$$

so that the Euler–Lagrange equation is

$$\frac{d[\dot{y}(1 + \dot{y}^2)^{1/2}(y_a - y)^{-1/2}]}{dx} - \frac{1}{2}(1 + \dot{y}^2)^{1/2}(y_a - y)^{-3/2} = 0$$

$$(C.3.1d)$$

Equation (C.3.1d) is invariant to the translation group $x' = x + \lambda$, $y' = y$, the invariants of which are y and $\dot{y}$. We can rewrite Eq. (C.3.1d) in terms of these variables and so reduce the order of the differential equation by dividing by $\dot{y}$:

$$\frac{d[\dot{y}(1 + \dot{y}^2)^{-1/2}(y_a - y)^{-1/2}]}{dy} - \frac{1}{2}(1 + \dot{y}^2)^{1/2}\frac{(y_a - y)^{-3/2}}{\dot{y}} = 0$$

$$(C.3.2a)$$

This last equation can be rewritten as

$$2[\dot{y}(1 + \dot{y}^2)^{-1/2}(y_a - y)^{-1/2}]\frac{d}{dy}[\dot{y}(1 + \dot{y}^2)^{-1/2}(y_a - y)^{-1/2}]$$

$$= (y_a - y)^{-2}$$

$$(C.3.2b)$$

where now we consider $\dot{y}$ to be a function of y. Integrating Eq. (C.3.2b), we now find

$$\dot{y}^2(1 + \dot{y}^2)^{-1}(y_a - y)^{-1} = (y_a - y)^{-1} - C^2 \qquad (C.3.3)$$

where C^2 is a constant of integration. Solving first for $\dot{y}$, we can finally write Eq. (C.3.3) as

$$dx = -\left[\frac{C^2(y_a - y)}{1 - C^2(y_a - y)}\right]^{1/2} dy \qquad (C.3.4)$$

We have taken the minus sign in the square root in Eq. (C.3.4) because when $dx > 0$, $dy < 0$. If we set $C^2(y_a - y) = \sin^2\theta$, we find after a short calculation

$$x = a + (2C^2)^{-1}[2\theta - \sin(2\theta)] \qquad (C.3.5a)$$

$$y = y_a - (2C^2)^{-1}[1 - \cos(2\theta)] \qquad (C.3.5b)$$

The value of C^2 as well as the value of θ_{max} are determined by the boundary condition $y = y_b$ at $x = b$. Equations (C.3.5) are the parametric equations of a cycloid.

The brachistochrone problem, proposed as a challenge problem in 1696 by John Bernoulli, was solved by him, his brother James, Leibniz, Newton and l'Hôpital. It opened the way for the development of a vast field known as the calculus of variations. The interested reader may find an excellent description in chapter IV of volume I of [Co-53]. The brief summary given above suffices for the purposes of this book.

Appendix D

Computation of Invariants and First Differential Invariants from the Transformation Equations

---

The modern style of referring to groups is in terms of their infinitesimal transformations (3.7.1) rather than in terms of their transformation equations (1.1.1). Indeed, in order to display the transformation equations when one knows the infinitesimal coefficients ξ and η one must integrate Eqs. (1.2.4). To calculate an invariant $u(x, y)$ and a first differential invariant $v(x, y, \dot{y})$, on which the application of Lie's reduction theorem depends, it is necessary to integrate the characteristic Eqs. (1.7.2b). But when the transformation equations are known explicitly: (1) an invariant can always be calculated by purely algebraic manipulation; and (2) a first differential invariant can always be calculated by differentiation and algebraic manipulation.

Proof:
(1) We begin by eliminating λ from the transformation equations (1.1.1a, b) to obtain a relation among x', y', x and y in which λ does not appear explicitly:

$$F(x', y', x, y) = 0 \qquad (D.1)$$

Now the relation (D.1) must hold for a fixed (x, y) as (x', y') varies over the entire orbit on which (x, y) lies. We show next that the function F can be rewritten as $G(u(x', y'), u(x, y))$, where $u(x, y)$ is an invariant, thus allowing us to identify $u(x, y)$ by purely algebraic means.

If we hold x and y fixed and let (x', y') move along an orbit O, we then find

$$F_{x'} \, dx' + F_{y'} \, dy' = 0 \tag{D.2}$$

where the infinitesimal vector (dx', dy') is tangent to the orbit O at (x', y'). The orbit O is also described by the equation

$$u(x', y') - c = 0 \tag{D.3}$$

where $c = u(x, y)$ (remember the point (x, y) is being held fixed). Thus

$$u_{x'} \, dx' + u_{y'} \, dy' = 0 \tag{D.4}$$

In order that Eqs. (D.2) and (D.4) have a nonzero solution for (dx', dy'), the determinant of the coefficients must vanish:

$$F_{x'} u_{y'} - F_{y'} u_{x'} = 0 \tag{D.5}$$

Eq. (D.5) is the condition that F and u be functionally dependent. Therefore,

$$F(x', y', x, y) = H(u(x', y'), x, y) \tag{D.6a}$$

Since we could have held (x', y') fixed and allowed (x, y) to vary over O, it is clear that

$$F(x', y', x, y) = G(u(x', y'), u(x, y)) \tag{D.6b}$$

as was to be proved.

(2) In the case of a first differential invariant, we must add Eq. (1.5.4) to the transformation Eqs. (1.1.1a, b). We denote the right-hand side of Eq. (1.5.4) henceforth by $P(x, y, \dot{y})$. The calculation of $P(x, y, \dot{y})$ involves only differentiation. Eliminating λ from Eqs. (1.5.4) and (1.1.1a) we find

$$F(x', \dot{y}', x, y, \dot{y}) = 0 \tag{D.7a}$$

while eliminating λ from Eqs. (1.5.4) and (1.1.1b) we find

$$G(y', \dot{y}', x, y, \dot{y}) = 0 \qquad (D.7b)$$

If we now eliminate $\dot{y}'$ between Eqs. (D.7a) and (D.7b) we obtain

$$K(x', y', x, y, \dot{y}) = 0 \qquad (D.8)$$

If we now fix x, y and $\dot{y}$ with their values at some specified point P on the orbit O and let (x', y') vary along O, we see by the same reasoning as before that

$$K(x', y', x, y, \dot{y}) = L(u(x', y'), x, y, \dot{y}) \qquad (D.9)$$

Solving $K = 0$ for u, we find

$$c = u(x', y') = M(x, y, \dot{y}) \qquad (D.10)$$

Since Eqs. (D.10) hold no matter what point P on O is originally chosen to be held fixed, we see that we have found a first differential invariant $M(x, y, \dot{y})$ by differentiation and algebraic manipulation.

Example: $x' = e^\lambda x, y' = (y^2 + \lambda)^{1/2}$, which extends to $\dot{y}' = y\dot{y}/[e^\lambda(y^2 + \lambda)^{1/2}]$. It follows straightforwardly then that $y'^2 = y^2 + \lambda = y^2 + \ln(x'/x)$ or $y'^2 - \ln x' = y^2 - \ln x$. The equation $\dot{y}' = y\dot{y}/[e^\lambda(y^2 + \lambda)^{1/2}]$ can be rewritten as $y'\dot{y}' = y\dot{y}/e^\lambda$ or $x'y'\dot{y}' = xy\dot{y}$.∎

Solutions to Problems

Chapter 1

1.1 We demonstrate the group property as follows:

$$x'' = G[G^{-1}(x') + \lambda'] = G[G^{-1}(G\{G^{-1}(x) + \lambda\}) + \lambda']$$
$$= G[G^{-1}(x) + \lambda + \lambda']$$
$$y'' = F[F^{-1}(y') - \lambda'] = F[F^{-1}(F\{F^{-1}(y) - \lambda\}) - \lambda']$$
$$= F[F^{-1}(y) - \lambda - \lambda']$$

The transformation with the group parameter $-\lambda$ is the inverse of the one with the group parameter λ; $G^{-1}(x) + \lambda = G^{-1}(x')$ so that $x = G[G^{-1}(x') - \lambda]$ and similarly for y. Finally, the value $\lambda = 0$ of the group parameter corresponds to the identity transformation.

1.2

(a) Let P: $(x + dx, y + dy)$ and Q: (x, y) be two neighboring points on a particular orbit corresponding to the values $\lambda + d\lambda$ and λ of the group parameter. Then $dx = \xi \, d\lambda$ and $dy = \eta \, d\lambda$ so that

$$\frac{df}{d\lambda} = \lim_{d\lambda \to 0} \frac{f(x + dx, y + dy) - f(x, y)}{d\lambda}$$
$$= \xi \frac{\partial f}{\partial x} + \eta \frac{\partial f}{\partial y}$$

(b)

$$\frac{d^2 f}{d\lambda^2} = \frac{d[\xi(\partial f/\partial x) + \eta(\partial f/\partial y)]}{d\lambda}$$

$$= \left(\xi\frac{\partial}{\partial x} + \eta\frac{\partial}{\partial y}\right)\left(\xi\frac{\partial f}{\partial x} + \eta\frac{\partial f}{\partial y}\right)$$

$$= \left(\xi\frac{\partial}{\partial x} + \eta\frac{\partial}{\partial y}\right)^2 f$$

(c)

$$\frac{d^n f}{d\lambda^n} = \left(\xi\frac{\partial}{\partial x} + \eta\frac{\partial}{\partial y}\right)^n f$$

$$f(x', y') = \sum_{n=0}^{\infty}\left(\xi\frac{\partial}{\partial x} + \eta\frac{\partial}{\partial y}\right)^n f(x, y)\frac{(\lambda' - \lambda)^n}{n!}$$

1.3 Abbreviate the linear differential operator $\xi(\partial/\partial x) + \eta(\partial/\partial y)$ by U. When $f(x, y) = x$, the Taylor series can be written

$$x' = \sum_{n=0}^{\infty}(U^n x)\frac{\lambda^n}{n!}$$

Now $Ux = -y$, $U^2 x = U(-y) = -x$, $U^3 x = y$, $U^4 x = x$, and cyclic repetition hereafter. Thus

$$x' = x - \lambda y - \left(\frac{\lambda^2}{2!}\right)x + \left(\frac{\lambda^3}{3!}\right)y + \left(\frac{\lambda^4}{4!}\right)x + \dots$$

$$= x\left(1 - \frac{\lambda^2}{2!} + \frac{\lambda^4}{4!} + \dots\right) - y\left(\lambda - \frac{\lambda^3}{3!} + \frac{\lambda^5}{5!} - \dots\right)$$

$$= x\cos\lambda - y\sin\lambda$$

and similarly for the equation for y'.

1.4

(a) The slope of the trajectories orthogonal to the orbits is $-\xi/\eta$. If P' and P are two neighboring points on an orbit, then if the image at P′ of the orthogonal trajectory through P is also an orthogonal trajectory, we must have (to first order in $d\lambda$)

$$-\frac{\xi'}{\eta'} = -\frac{\xi}{\eta} + d\lambda\,\eta_1 = -\frac{\xi}{\eta} + d\lambda\left(\frac{d\eta}{dx} + \frac{\xi}{\eta}\frac{d\xi}{dx}\right) \quad (1)$$

Now,

$$\xi' = \xi + d\lambda\,U\xi \quad \text{and} \quad \eta' = \eta + d\lambda\,U\eta$$

where U is again the linear differential operator $\xi(\partial/\partial x) + \eta(\partial/\partial y)$. Thus

$$-\frac{\xi'}{\eta'} = -\frac{\xi}{\eta}\left(1 + \frac{d\lambda(U\xi)}{\xi} - \frac{d\lambda(U\eta)}{\eta}\right) \quad (2)$$

Equating the right-hand sides of Eqs. (1) and (2) and rearranging, we find

$$2\frac{\xi}{\eta}\frac{\partial\xi}{\partial x} + \left(1 - \frac{\xi^2}{\eta^2}\right)\frac{\partial\xi}{\partial y} = 2\frac{\xi}{\eta}\frac{\partial\eta}{\partial y} - \left(1 - \frac{\xi^2}{\eta^2}\right)\frac{\partial\eta}{\partial x} \quad (3)$$

(b) If ξ is a function of y only and η is a function of x only, Eq. (3) reduces to the equation $\partial\xi/\partial y = -\partial\eta/\partial x$ *in which the variables are separated*. Integrating, we find that most generally $\xi = ay + b_1$ and $\eta = -ax + b_2$, where a, b_1 and b_2 are constants, each choice of which determines a different group. The orbits of any of these groups must take the form $(a/2)(x^2 + y^2) + b_1y - b_2x = c$, where c is the parameter labelling the orbits. These orbits are concentric circles; their orthogonal trajectories are therefore the family of rays emanating from the common center of the orbits.

(c) If ξ is a function of x only and η is a function of y only, Eq. (3) reduces to the equation $\partial\xi/\partial x = \partial\eta/\partial y$ *in which the variables are again separated*. Integrating, we find that most generally $\xi = ax + b_1$ and $\eta = ay + b_2$, where a, b_1 and b_2 are constants, each choice of which determines a different group. The orbits of any of these groups must take the form $ay + b_2 = c(ax + b_1)$, where c is the parameter labelling the orbits. The orbits are thus a family of rays emanating from the point $(-b_1/a, -b_2/a)$ and their orthogonal trajectories are a family of concentric circles.

1.5 Suppose a parametric representation of the orbits is $\phi(x, y) = c_1$ and that of the orthogonal trajectories is $\psi(x, y) = c_2$. The values of the functions ϕ and ψ together determine the position of a point (x, y). We can imagine moving along an orbit then by keeping the value of ϕ fixed and changing the value of ψ. This suggests that the transformations

$$\phi(x', y') = \phi(x, y) \quad \text{and} \quad \psi(x', y') = \psi(x, y) + \lambda \quad (1)$$

may form a group. This supposition is easily verified:

$$\phi(x'', y'') = \phi(x', y') = \phi(x, y) \quad (2a)$$

$$\psi(x'', y'') = \psi(x', y') + \lambda' = \psi(x, y) + \lambda + \lambda' \quad (2b)$$

Equations (2) verify the group property. The identity transformation is that for which $\lambda = 0$ and the transformation corresponding to $-\lambda$ is inverse to that corresponding to λ.

The reader should note that nowhere have we used the hypothesis that $\psi = c$ denotes the orthogonal trajectories of the orbits. Thus it should be clear that any pair of functionally independent functions $\phi(x, y)$ and $\psi(x, y)$ can determine an additive group like (1). The functional independence is necessary in order that the equations $\xi\phi_x + \eta\phi_y = 0$ and $\xi\psi_x + \eta\psi_y = 1$ obtained by differentiating Eqs. (1) with respect to λ and setting $\lambda = 0$, have a nontrivial solution for ξ and η.

1.6

(a) In this problem, Eqs. (1.2.4) are $x \, dx = -2y \, dy = d\lambda$. The first pair of these equations can be integrated to give $x^2/2 + y^2 = c$; the second pair can be integrated to give $y^2 + \lambda = a$. Here c and a are constants of integration. Combining the last two results we find $x^2 = 2[c - a + \lambda]$. If we take $\lambda = 0$ to be the identity transformation, we then find for the transformation equations $x' = (x^2 + 2\lambda)^{1/2}$, $y' = (y^2 - \lambda)^{1/2}$.

(b) Equations (1.2.4) are now $dx/y = dy/x = d\lambda$. The first pair can be integrated to yield $x^2 - y^2 = c^2$. By substituting $x = (y^2 + c^2)^{1/2}$, we can integrate the second pair to give $y = c\sinh(\lambda + a)$. These two results can be combined to give $x = c\cosh(\lambda+a)$. If we take $x = c\cosh a$, $y = c\sinh a$ and $\lambda = 0$ to be the identity transformation, we find $x' = x\cosh\lambda + y\sinh\lambda$, $y' = y\cosh\lambda + x\sinh\lambda$ for the transformation equations.

(c) Equations (1.2.4) are now $dx/x^2 = dy/xy = d\lambda$. We find from the first pair that $y = cx$. Substituting this value of y into the second pair, we can integrate the latter to give $1/x + \lambda = a$. According to the first of these results, $y'/x' = y/x$. If we again take $\lambda = 0$ to be the identity transformation, the second result gives, after rearrangement, $x' = x/(1 - \lambda x)$. Combining it with the first result we find $y' = y/(1 - \lambda x)$.

(d) Equations (1.2.4) are now $dx/(x - y) = dy/(x + y) = d\lambda$. The first equation of this pair can be cleared of fractions to give the homogeneous equation $(x + y)\,dx - (x - y)\,dy = 0$, which has the integrating factor $1/(x^2 + y^2)$. (If the reader is not familiar with integrating factors, he may delay reading the rest of this solution until after the next section, section 2.1). Then, if the family of solutions is represented as $\phi(x, y) = c$, we must have

$$\frac{\partial\phi}{\partial x} = \frac{x + y}{x^2 + y^2} \quad \text{and} \quad \frac{\partial\phi}{\partial y} = \frac{y - x}{x^2 + y^2} \qquad (1)$$

The first of these equations can be integrated to give

$$\phi = \ln(x^2 + y^2)^{1/2} + \arctan\left(\frac{x}{y}\right) + F(y) \qquad (2)$$

where the arbitrary function $F(y)$ acts as a constant of integration. Inserting this value of ϕ into the second equation of the pair Eq. (1), we find by comparing the left- and right-hand sides that $dF/dy = 0$ so that F is a constant. We can absorb the constant F into the constant c, so that

the orbits are given by constant values of the function
$\phi = \ln(x^2 + y^2)^{1/2} + \arctan(x/y)$.

The function ϕ is greatly simplified by the introduction of polar coordinates, which is strongly hinted at by its form. If we set $x = r\cos\theta$ and $y = r\sin\theta$, we find $\phi = \ln r + (\pi/2 - \theta)$. Thus $\ln r' + (\pi/2 - \theta') = \ln r + (\pi/2 - \theta)$ or $r'/r = \exp(\theta' - \theta)$. This strongly suggests that the group transformations have the form $\theta' = \theta + \lambda, r' = re^\lambda$.

To verify this supposition, we use the chain rule to calculate the coefficients ξ and η:

$$\xi = \left(\frac{\partial x'}{\partial \lambda}\right)_{\lambda=0} = \left(\frac{\partial x'}{\partial r'}\frac{\partial r'}{\partial \lambda} + \frac{\partial x'}{\partial \theta'}\frac{\partial \theta'}{\partial \lambda}\right)_{\lambda=0}$$
$$= r\cos\theta - r\sin\theta = x - y$$

$$\eta = \left(\frac{\partial y'}{\partial \lambda}\right)_{\lambda=0} = \left(\frac{\partial y'}{\partial r'}\frac{\partial r'}{\partial \lambda} + \frac{\partial y'}{\partial \theta'}\frac{\partial \theta'}{\partial \lambda}\right)_{\lambda=0}$$
$$= r\sin\theta + r\cos\theta = y + x$$

Chapter 2

2.1 A suitable group is the stretching group $y' = y/\lambda, x' = \lambda x$ for which $\xi = x$ and $\eta = -y$. By rewriting the differential equation in the form of Eq. (2.1.2), we find that $M = 4x^3y^2 + 2x$ and $N = 2x^4y$. Lie's integrating factor is then $(2x^4y^2 + 2x^2)^{-1}$. If the integral curves are parametrized as in Eq. (1.4.5), then

$$\psi_x = \frac{4x^3y^2 + 2x}{2x^4y^2 + 2x^2} \tag{1a}$$

$$\psi_y = \frac{2x^4y}{2x^4y^2 + 2x^2} = \frac{x^2y}{x^2y^2 + 1} \tag{1b}$$

Integrating Eq. (1b), we find $\psi = (1/2)\ln(x^2y^2 + 1) + F(x)$, where the as yet unknown function $F(x)$ serves as the constant of integration. Differentiating this expression for ψ partially with respect to x and comparing the result with Eq. (1a), we find that

$dF/dx = 1/x$ so that $F = \ln x + a$, where a is a constant of integration. Absorbing a into the constant value of ψ, we find the integral curves of the differential equation are given by constant values of the function $\psi = (1/2)\ln(x^4 y^2 + x^2)$ or equivalently by constant values of the function $x^4 y^2 + x^2 = e^{2\psi} \equiv c$. Solved explicitly for the variable y, this expression becomes $y = (c - x^2)^{1/2}/x^2$.

2.2 In this problem, in contrast to problem 2.1, the group is not easily found by inspection. Let us look for groups for which the quantity $\ln y + x^2$ is a group invariant. Let us also note that if any of these groups leaves the differential equation invariant, then $\dot{y}/xy$ must be a first differential invariant of that group. According to Eqs. (1.3.2) and (1.7.2a), the invariance of these two expressions leads to the conditions

$$\frac{\eta}{y} = -2x\xi \tag{1a}$$

$$\eta_1 = \dot{y}\left(\frac{\xi}{x} + \frac{\eta}{y}\right) \tag{1b}$$

Using Eq. (1.5.5), we can write Eq. (1b) as

$$\frac{\partial \eta}{\partial x} + \dot{y}\left(\frac{\partial \eta}{\partial y} - \frac{\partial \xi}{\partial x}\right) - \dot{y}^2\frac{\partial \xi}{\partial y} = \dot{y}\left(\frac{\xi}{x} + \frac{\eta}{y}\right) \tag{2}$$

*Since in the determination of a group having the specified invariants, no connection is implied among x, y and $\dot{y}$, Eq. (2) is an **identity** in $\dot{y}$, i.e. holds for all values of $\dot{y}$.* Thus we must have

$$\frac{\partial \eta}{\partial x} = 0 \tag{3a}$$

$$\frac{\partial \eta}{\partial y} - \frac{\partial \xi}{\partial x} = \frac{\xi}{x} + \frac{\eta}{y} \tag{3b}$$

$$\frac{\partial \xi}{\partial y} = 0 \tag{3c}$$

According to Eq. (3c) ξ must be a function only of x and according to Eq. (3a) η must be a function only of y. Thus we see at once from Eq. (1*a*) that $\xi = c/(2x)$ and $\eta = -cy$, where c is a (separation) constant that we can choose at will; we make the choice $c = 1$ for convenience. That these expressions for ξ and η satisfy Eq. (3*b*) can easily be verified. Now we are in a position to calculate Lie's integrating factor. [Although not required for finishing the problem, the transformation equations for the group we have just found are $y' = ye^{-\lambda}$, $x' = (x^2 + \lambda)^{1/2}$.]

The coefficients M and N in the form (2.1.2) of the differential equation are $M = xy(1 + \ln y + x^2)$ and $N = -1$. If the integral curves are parametrized as in Eq. (1.4.5), then

$$\psi_x = 2x \frac{1 + \ln y + x^2}{3 + \ln y + x^2} \tag{4a}$$

$$\psi_y = \frac{-2}{y(3 + \ln y + x^2)} \tag{4b}$$

If we integrate Eq. (4*b*), we find $\psi = -2\ln(3 + \ln y + x^2) + F(x)$, where $F(x)$ serves as an arbitrary constant of integration. If we differentiate this last expression partially with respect to x and compare with Eq. (4*a*), we find $dF/dx = 2x$ so that $F = x^2 + c$, a constant. Absorbing this constant into the constant value of ψ and rewriting, we find that the integral curves are given by constant values of the function

$$e^{\psi} = \frac{\exp(x^2)}{(3 + \ln y + x^2)^2} \tag{5}$$

2.3 Writing the differential equation in the form (2.1.2), we see that $M = P(x)y - Q(x)$ and $N = 1$. Then according to the converse of Lie's theorem,

$$\xi[P(x)y - Q(x)] + \eta = \exp\left(-\int_0^x P(z)\,dz\right) \tag{1}$$

(a) Try $\xi = 0$. Then $\eta = \exp(-\int_0^x P(z)\,dz)$ so that Eqs. (1.2.4) become

$$\frac{dx}{0} = \frac{dy}{\exp(-\int_0^x P(z)\,dz)} = d\lambda \tag{2}$$

Equations (2) can be integrated to give $x = c$, $y = \lambda \exp(-\int_0^c P(z)\,dz) + a$, where c and a are constants of integration. If we choose $\lambda = 0$ to denote the identity transformation and take (c, a) to be the coordinates of the source point, we obtain the transformation equations

$$x' = x \qquad y' = y + \lambda \exp\left(-\int_0^x P(z)\,dz\right)$$

(b) Try $\eta = 0$. Then $\xi = \exp(-\int_0^x P(z)\,dz)/(Py - Q)$ and Eqs. (1.2.4) become

$$dx(Py - Q)\exp\left(\int_0^x P(z)\,dz\right) = \frac{dy}{0} = d\lambda$$

which can be integrated to give

$$y = c$$

$$\lambda + a = y\int_0^x P(w)\exp\left(\int_0^w P(z)\,dz\right)dw$$

$$- \int_0^x Q(w)\exp\left(\int_0^w P(z)\,dz\right)dw$$

These equations lead to the transformation equations

$$y = y'$$

$$\lambda = y\int_x^{x'} P(w)\exp\left(\int_0^w P(z)\,dz\right)dw$$

$$- \int_x^{x'} Q(w)\exp\left(\int_0^w P(z)\,dz\right)dw$$

2.4 The translation group $x' = x + \lambda$, $y' = y - (a/b)\lambda$ leaves both sides of the differential equation $\dot{y} = ax + by + c$ unchanged.

Thus $\xi = 1$ and $\eta = -a/b$. Since $M = ax + by + c$ and $N = -1$, Lie's integrating factor is $1/(ax + by + c + a/b)$. If the integral curves are parametrized as in Eq. (1.4.5), then

$$\psi_x = \frac{ax + by + c}{ax + by + c + a/b}$$

$$= 1 - \frac{a/b}{ax + by + c + a/b} \tag{1a}$$

$$\psi_y = \frac{-1}{ax + by + c + a/b} \tag{1b}$$

If we integrate Eq. (1a) with respect to x and Eq. (1b) with respect to y, we find, respectively,

$$\psi = x - \frac{1}{b}\ln\left(ax + by + c + \frac{a}{b}\right) + f(y) \tag{2a}$$

$$\psi = -\frac{1}{b}\ln\left(ax + by + c + \frac{a}{b}\right) + g(x) \tag{2b}$$

where the arbitrary functions $f(y)$ and $g(x)$ serve as constants of integration. Comparing Eqs. (2a) and (2b) we see that

$$\psi = x - \frac{1}{b}\ln\left(ax + by + c + \frac{a}{b}\right)$$

2.5 In order to show that the differential equation

$$v(x, y, \dot{y}) = (\ln y - e^x)\exp(x + e^x) - \dot{y} = 0 \tag{1}$$

is invariant to the group

$$x' = \ln(e^x + \lambda) \qquad y' = \lambda y \tag{2}$$

we show that $v(x, y, \dot{y})$ satisfies Eq. (1.7.2a) when $v(x, y, \dot{y}) = 0$. By direct calculation we find the infinitesimal coefficients of the group to be $\xi = e^{-x}$ and $\eta = y$; from Eq. (1.5.5) we find that $\eta_1 = \dot{y}(1 + e^{-x})$. A straightforward calculation now shows that

$$\xi v_x = \exp(x + e^x)[(1 + e^{-x})(\ln y - e^x) - 1] \tag{3a}$$

$$\eta v_y = \exp(x + e^x) \tag{3b}$$

$$\eta_1 v_{\dot{y}} = -\dot{y}(1 + e^{-x})$$

$$= -\exp(x + e^x)(\ln y - e^x)(1 + e^{-x}) \tag{3c}$$

where we have substituted for $\dot{y}$ from Eq. (1). Thus Eq. (1.7.2a) is satisfied.

To calculate the new variables $\underline{x} = F(x, y)$ and $\underline{y} = G(x, y)$, we must solve Eqs. (2.5.6), for which the characteristic equations are, respectively,

$$e^x\,dx = \frac{dy}{y} \qquad \text{and} \qquad e^x\,dx = \frac{dy}{y} = dG \qquad (4)$$

From these equations we find that F and $(G - \ln y)$ are arbitrary functions of $\ln y - e^x$. Therefore, a simple choice of new variables is

$$\underline{x} = \ln y - e^x \qquad \text{and} \qquad \underline{y} = \ln y \qquad (5)$$

When inverted, these equations read

$$x = \ln(\ln \underline{y} - \underline{x}) \qquad \text{and} \qquad y = \exp(\underline{y}) \qquad (6)$$

From (6), we determine after a short calculation that

$$\dot{y} = \underline{\dot{y}}(\underline{y} - \underline{x})\exp(\underline{y})/(\underline{y} - 1) \qquad (7)$$

Substituting (6) and (7) into the differential Eq. (1), we finally find the separated equation

$$\underline{\dot{y}} = -\frac{\underline{x}}{e^{\underline{x}} - \underline{x}} \qquad (8)$$

2.6 If we eliminate by differentiation the constant c from the parametric representation $\psi(x, y) = c$ of the family of integral curves, we find the differential equation $u(x, y, \dot{y}) = \dot{y} + \psi_x/\psi_y = 0$. With this definition of u, the left-hand side of Eq. (1.7.2a) becomes (after some rearrangement)

$$\xi(\psi_{xx}\psi_y - \psi_{xy}\psi_x) + \eta(\psi_{xy}\psi_y - \psi_{yy}\psi_x)$$
$$+ \psi_y^2\left[\eta_x - \frac{\psi_x}{\psi_y}(\eta_y - \xi_x) - \left(\frac{\psi_x}{\psi_y}\right)^2\xi_y\right] \qquad (1)$$

which can be written

$$\psi_y\frac{\partial(\xi\psi_x + \eta\psi_y)}{\partial x} - \psi_x\frac{\partial(\xi\psi_x + \eta\psi_y)}{\partial y} \qquad (2)$$

In view of Eq. (1.4.9), the two partial derivatives in Eq. (2) vanish individually, which shows that expressions (2) and (1) both vanish and thus completes the proof.

Conversely, if the quantity in (1) vanishes then so does the quantity in (2), which then becomes the following partial differential equation for the function $\varphi \equiv \xi \psi_x + \eta \psi_y$: $\psi_y \partial \varphi / \partial x - \psi_x \partial \varphi / \partial y = 0$. The characteristic equations are then $dx/\psi_y = -dy/\psi_x = d\varphi/0$. The functions ψ and φ are two independent integrals of these equations, so the most general form for φ is $F(\psi)$, where F is an arbitrary function. But this means that ψ satisfies Eq. (1.4.4), so that the family given by $\psi = c$ is invariant.

2.7 The stretching group $y' = \lambda^{-m/n}y, x' = \lambda x$, for which $\xi = x$ and $\eta = -my/n$, suffices. Lie's integrating factor is $\mu^{-1} = xy[(a - mb/n) + (\alpha - m\beta/n)x^m y^n]$ so that the general solution is $\mu/\nu = c$ as long as the ratio μ/ν is not constant.

2.8 For the rotation group $\xi = -y$ and $\eta = x$, so that $\eta_1 = 1 + \dot{y}^2$. Equations (1.7.2b) then become $-dx/y = dy/x = d\dot{y}/(1 + \dot{y}^2)$. From the first equality we find that $x^2 + y^2 = c^2$, where c is a constant of integration. Substituting $x = (c^2 - y^2)^{1/2}$, we find from the second equality the integral $\arctan \dot{y} - \arcsin(y/c)$. Thus the most general solution of Eq. (1.7.2a) is

$$\arctan \dot{y} - \arcsin\left(\frac{y}{[x^2 + y^2]^{1/2}}\right) = G(x^2 + y^2) \qquad (1)$$

where G is an arbitrary function. By moving the arcsin to the right-hand side of Eq. (1) and taking the tangent of both sides, we find

$$\dot{y} = \frac{y + xF}{x - Fy} \qquad (2)$$

where $F = \tan G$ is also an arbitrary function of $x^2 + y^2$. Eq. (2) is identical with Davis' equation as was to be shown. Since $M = y + xF$ and $N = x - Fy$, Lie's integrating factor $\mu = (\xi M + \eta N)^{-1} = -(x^2 + y^2)^{-1}$. Because any constant multiple of an integrating factor is also an integrating factor, $(x^2 + y^2)^{-1}$ is an integrating factor.

2.9 Chrystal's equation is invariant to the stretching group $y' = \lambda^2 y, x' = \lambda x$ an invariant of which is y/x^2. Since the singular solution is an invariant of all groups which leave the differential equation invariant, it must have the form $y = Dx^2$, where D is a constant yet to be determined. Substituting this form into Chrystal's equation, we find that D must satisfy the quadratic equation $4D^2 + (2A + B)D + C = 0$. In order that there be real solutions for D, the discriminant $4A^2 + 4AB + B^2 - 16C$ must be ≥ 0.

2.10

(a) Let θ be the angle between the x-axis and the normal PQ to C_1 at the point P: (x, y), and let δ be the constant length along this normal to the point Q: (x', y') of C_2 (draw a sketch!). The coordinates of Q are then

$$y' = y + \delta \sin \theta \qquad (1a)$$
$$x' = x + \delta \cos \theta \qquad (1b)$$

Now if $P_1: (x + dx, y + dy)$ is a point adjacent to P on C_1 and $Q_1: (x' + dx', y' + dy')$ is the point at which the normal to C_1 at P_1 intersects C_2, then

$$y' + dy' = y + dy + \delta \sin(\theta + d\theta) \, d\theta \qquad (2a)$$
$$x' + dx' = x + dx + \delta \cos(\theta + d\theta) \, d\theta \qquad (2b)$$

Thus

$$dy' = dy + \delta \cos \theta \qquad (3a)$$
$$dx' = dx - \delta \sin \theta \qquad (3b)$$

so that

$$\frac{dy'}{dx'} = \frac{dy}{dx} \frac{[1 + \delta \cos \theta (d\theta/dx)/(dy/dx)]}{[1 - \delta \sin \theta (d\theta/dx)]} \qquad (4)$$

Now $\tan \theta = -1/(dy/dx)$ since the slope of the normal is the negative reciprocal of the slope of the tangent. Substituting this value of dy/dx into Eq. (4), we find that $(dy'/dx')_Q = (dy/dx)_P$. This means that curve C_2 is perpendicular at point Q to the normal to curve C_1 at P, as was to be proved.

(b) From the result of part (a) it is clear that the orthogonal trajectories of the family F are a family of straight lines. The images under the transformations T_λ of the point (x, y) all lie on the same one of these straight lines as (x, y) itself (call this line L). The image (x', y') under the transformation T_λ lies at a distance λ from (x, y) along L. The image (x'', y'') of (x', y') under T_μ lies a distance μ from (x', y') along L as well. So (x'', y'') lies on L a distance $\lambda + \mu$ from (x, y). In other words, $T_\mu T_\lambda = T_{\mu+\lambda}$, which is the group property. Similarly, we see that $T_{-\lambda}$ is inverse to T_λ and T_0 is the identity transformation. Thus the collection $\{T_\lambda\}$ forms a group G.

(c) If the differential equation of the family F is $M(x, y)\,dx + N(x, y)\,dy = 0$, then $\tan\theta = -1/(dy/dx) = N/M$ so that $\cos\theta = M/(M^2 + N^2)^{1/2}$ and $\sin\theta = N/(M^2 + N^2)^{1/2}$. If we insert these values into Eqs. (1a, b) and replace δ by λ, we have

$$y' = y + \frac{\lambda N}{(M^2 + N^2)^{1/2}} \tag{5a}$$

$$x' = x + \frac{\lambda M}{(M^2 + N^2)^{1/2}} \tag{5b}$$

Differentiating each of these equations with respect to λ, we find

$$\eta = \frac{N}{(M^2 + N^2)^{1/2}} \tag{6a}$$

$$\xi = \frac{M}{(M^2 + N^2)^{1/2}} \tag{6b}$$

Then Lie's integrating factor becomes

$$\mu = (M^2 + N^2)^{-1/2} \tag{7}$$

(d) Conversely, if Eq. (7) is an integrating factor, then

$$\psi_x = \mu M = \frac{M}{(M^2 + N^2)^{1/2}} \tag{8a}$$

$$\psi_y = \mu N = \frac{N}{(M^2 + N^2)^{1/2}} \tag{8b}$$

so that $|\nabla\psi| = (\psi_x^2 + \psi_y^2)^{1/2} = 1$. Since the gradient vector $\nabla\psi$ is always normal to the level curves $\psi(x, y) = c$, in this example it is a unit vector in the normal direction.

Now if P: (x, y) is a point on the level curve $C_1: \psi(x, y) = c$ and Q: $(x + dx, y + dy)$ is a nearby point on the level curve $C_2: \psi(x + dx, y + dy) = c + dc$, then

$$\psi_x \, dx + \psi_y \, dy = dc \tag{9}$$

If the vector (dx, dy) also points in the direction of $\nabla\psi = (\psi_x, \psi_y)$, i.e. if the point Q is the intersection of the normal to curve C_1 at P with the curve C_2, then

$$[(dx)^2 + (dy)^2]^{1/2} = dc \tag{10}$$

so that the normal distance from C_1 to C_2 has the constant value dc. Thus C_1 and C_2 are parallel.

2.11

(a) When $\mu = e^x$, the partial differential equation $(\mu M)_y = (\mu N)_x$ becomes $M_y = N + N_x$, which straightforward computation shows is satisfied when $M = 3x^2y + y^3 + 6xy$, $N = 3(x^2 + y^2)$.

(b) Since $\xi M + \eta N = \mu^{-1}$, when $\xi = 0$, $\eta = \mu^{-1}/N = e^{-x}/3(x^2 + y^2)$.

(c) The transformation equations of the group are found by solving Eqs. (1.2.4), which in the present problem take the form

$$dx/0 = 3e^x(x^2 + y^2) \, dy = d\lambda \tag{1}$$

Two integrals of these equations are

$$x = a \qquad 3e^a\left(a^2y + \frac{y^3}{3}\right) = \lambda + b \tag{2}$$

where a and b are constants of integration. If we let $\lambda = 0$ denote the identity transformation, Eqs. (2) are equivalent to the transformation equations

$$x' = x \qquad 3e^x\left(x^2y' + \frac{y'^3}{3}\right) = \lambda + 3e^x\left(x^2y + \frac{y^3}{3}\right) \tag{3}$$

2.12

(a) If we differentiate the differential equation

$$y = x\dot{y} + (1 + \dot{y}^2)^{1/2} \tag{1}$$

with respect to the group parameter λ and set $\lambda = \lambda_0$, the group parameter corresponding to the identity transformation, we find

$$\eta = \xi\dot{y} + x\eta_1 + \dot{y}\eta_1(1 + \dot{y}^2)^{-1/2} \tag{2}$$

When $\xi = y$ and $\eta = -x$, it follows from Eq. (1.5.5b) that $\eta_1 = -(1 + \dot{y}^2)$. These values of ξ, η and η_1 satisfy Eq. (2) when x, y and $\dot{y}$ satisfy Eq. (1), which shows that the differential Eq. (1) is invariant to the rotation group.

(b) Since the envelope is an invariant curve of any group that leaves Eq. (1) invariant, it must be an invariant curve of the rotation group. Such curves obey the differential equation $dx/\xi = dy/\eta$ or $dx/y = -dy/x$. This last equation can be integrated immediately to give $x^2 + y^2 = c^2$, where c is a constant of integration. The envelope must also satisfy Eq. (1); substitution of the solution $x^2 + y^2 = c^2$ into Eq. (1) leads to the value $c = 1$.

2.13 The differential equation $\dot{y}^2 - 8x^3\dot{y} + 16x^2y = 0$ is invariant to the stretching group $y' = \lambda^4 y$, $x' = \lambda x$, the invariant curves of which have the form $y = cx^4$ where c is a constant of integration. The envelope, which must be an invariant curve, also satisfies the differential equation. Substituting $y = cx^4$ into the differential equation, one finds that $c = 1$.

2.14

(a) If we rewrite the differential equation $(2x^2y - x^3 - y)\,dx + x\,dy = 0$ in the form $\dot{y} + (2x - 1/x)y = x^2$, we see that $\exp(x^2)/x$ is an integrating factor.

(b) Using this integrating factor, the differential equation can be written in the form $d[y\exp(x^2)/x]/dx = x\exp(x^2)$. We see, therefore, that the substitution $y = ux\exp(-x^2)$ separates the variables u and x.

(c) If we identify x with $\underline{x} = F(x, y)$ and u with $\underline{y} = G(x, y)$ in Eqs. (2.5.6), we find that $\xi = 0$ and $\eta = x\exp(-x^2)$.

(d) If we differentiate the differential equation in the form $\dot{y} + (2x - 1/x)y = x^2$ with respect to the group parameter λ and set $\lambda = \lambda_0$, the group parameter corresponding to the identity transformation, we find $\eta_1 + (2\xi + \xi/x^2)y + (2x - 1/x)\eta = 2\xi x$. When $\xi = 0$ this relation becomes $d\eta/dx + (2x - 1/x)\eta = 0$; here we have used Eq. (1.5.5) for η_1. This last differential equation can be solved to give $\eta = Cx\exp(-x^2)$, where C is a constant of integration. The constant C can be absorbed into the group parameter λ (consider Eqs. (1.2.4)), so we have the same result for η as in part (c).

2.15

(a) If we differentiate Eqs. ($1a$, b) of the problem statement with respect to the group parameter λ and set $\lambda = \lambda_0$, the group parameter corresponding to the identity transformation, we find

$$\xi u_x + \eta u_y = 0 \quad \text{and} \quad \xi v_x + \eta v_y = 1 \tag{1}$$

so that u and v satisfy Eqs. (2.5.6) and thus can be identified as new variables that separate when substituted into a first-order differential equation invariant under G.

(b) We resort again to the method of problem 2.2. We look for a group for which $y^2 - \ln x$ is an invariant and for which $y\dot{y}\exp(y^2)$ is therefore a first differential invariant. As in the solution of problem 2.2, the invariance of these two expressions leads to the conditions

$$\xi = 2xy\eta \qquad (2a)$$

$$\eta_x + \dot{y}\left[\eta_y - \xi_x + \eta\left(2y + \frac{1}{y}\right)\right] - \dot{y}^2\xi_y = 0 \qquad (2b)$$

Since in the determination of a group having the specified invariants, no connection is implied among x, y and $\dot{y}$, Eq. (2) is an identity in $\dot{y}$, i.e. holds for all values of $\dot{y}$. Thus we must have

$$\eta_x = 0 \qquad (3a)$$

$$\xi_y = 0 \qquad (3b)$$

$$\eta_y - \xi_x + \eta\left(2y + \frac{1}{y}\right) = 0 \qquad (3c)$$

According to Eq. (3a), η is a function of y only. Then from Eqs. (3b) and (2a) we find that $(y\eta)_y = 0$ so that $\eta = c/y$, where c is a constant. Then $\xi = 2cx$. Finally, these expressions for ξ and η satisfy Eq. (3c).

Now Eqs. (1.2.4) can be written

$$\frac{dx}{x} = 2y\,dy = d\lambda \qquad (4)$$

where we have subsumed a constant factor $2c$ in the group parameter λ. Equations (4) can be integrated to give the redundant transformation equations

$$y'^2 - \ln x' = y^2 - \ln x \qquad (5a)$$

$$\ln x' = \ln x + \lambda \qquad (5b)$$

$$y'^2 = y^2 + \lambda \qquad (5c)$$

If we choose

$$u = y^2 - \ln x \quad \text{and} \quad v = \ln x \qquad (6a)$$

which when inverted give

$$y = (u + v)^{1/2} \quad \text{and} \quad x = e^v \qquad (6b)$$

then Eqs. (1*a, b*) in the problem statement are satisfied. To Eqs. (6*b*) we add the equation

$$\dot{y} = (u + v)^{-1/2} e^{-v} \frac{1 + \dot{u}}{2} \qquad \text{where} \qquad \dot{u} \equiv \frac{du}{dv} \qquad (7)$$

Substituting Eqs. (6*b*) and (7) in differential equation (2) of the problem statement, we find

$$du + (2u^{1/2}e^{-u} + 1) \, dv = 0 \qquad (8)$$

in which the variables u and v are clearly separable.

2.16

(a) When we make the substitution $y = u^{1/(1-n)}$ in the Bernoulli equation, we obtain the following linear equation

$$\frac{du}{dx} + (1 - n)P(x)u = (1 - n)Q(x) \qquad (1)$$

An integrating factor for this equation is $\mu = \exp[(1 - n) \int P \, dx]$; multiplying Eq. (1) by μ yields

$$\frac{d\{u \exp[(1 - n) \int P \, dx]\}}{dx} = (1-n)Q(x) \exp\left[(1-n) \int P \, dx\right] \qquad (2)$$

Thus if we make the substitution

$$v = u \exp\left[(1 - n) \int P dx\right] = y^{1-n} \exp\left[(1 - n) \int P \, dx\right] \qquad (3)$$

for y in the Bernoulli equation, the new variables x and v separate.

(b) If we use the variables x and v for F and G, respectively, in Eqs. (2.5.6), we find that

$$\xi = 0 \qquad \text{and} \qquad \eta = \frac{y^n \exp[(n - 1) \int P dx]}{1 - n} \qquad (4)$$

(c) If we differentiate the Bernoulli equation with respect to the group parameter λ and set $\lambda = \lambda_0$, the group parameter corresponding to the identity transformation, we find, taking into account that $\xi = 0$,

$$\eta_1 + P\eta = nQy^{n-1}\eta \qquad (5)$$

Also when $\xi = 0$,

$$\eta_1 = \eta_x + \dot{y}\eta_y = \eta_x + (Qy^n - Py)\eta_y \qquad (6)$$

so that Eq. (5) may be written

$$\eta_x + (Qy^n - Py)\eta_y + P\eta = nQy^{n-1}\eta \qquad (7)$$

Direct calculation now shows that η given in Eq. (4) satisfies Eq. (7).

2.17   If we write the differential equation $\dot{y} = y$ in the form (2.1.2) then $M = y$ and $N = -1$. The integral curves $y = ce^x$ may be written in the form $\psi = c$, where $\psi = ye^{-x}$. Now the differential equation is invariant to the extended group $x' = x$, $y' = \lambda y$, $\dot{y}' = \lambda \dot{y}$. Thus $\xi = 0$, $\eta = y$ and $\eta_1 = \dot{y}$. Therefore $\mu = (\xi M + \eta N)^{-1} = -1/y$ so that according to Eq. (2.3.3), $\nu = -F(ye^{-x})/y$.

We would like to find a group that leaves the curve $y = c_0 e^x$ invariant. We do this by finding an integrating factor whose reciprocal vanishes on the curve $y = c_0 e^x$, as explained in section 2.3. Try $F(z) = (z - c_0)^{-1}$. Then $\nu^{-1} = y(c_0 - ye^{-x})$ and this must equal $\xi M + \eta N = \xi y - \eta$. If we take $\xi = c_0$, then $\eta = y^2 e^{-x}$ and $\eta_1 = (2y\dot{y} - y^2)e^{-x}$. (This group does in fact leave the differential equation invariant, the condition for this being $\eta = \eta_1$ when $\dot{y} = y$.)

The finite form of this group is found by integrating Eqs. (1.2.4), which now take the form $dx/c_0 = dy/y^2 e^{-x} = d\lambda$. The first equality yields $-e^{-x} + c_0/y = a'$, the second $x = c_0\lambda + a''$, where a' and a'' are constants of integration. If we take $\lambda = 0$ to correspond to the identity transformation and choose $a'' = x_0$ and $a' = -e^{-x_0} + c_0/y_0$, we find that the finite group can be written $x' = x + c_0\lambda$, $y' = c_0\{c_0/y + [\exp(-\lambda c_0) - 1]e^{-x}\}^{-1}$. When $y = ce^x$, $y' = \exp(-\lambda c_0)[c_0/c + \exp(-\lambda c_0) - 1]^{-1}e^{x'}$, so that $c' = c_0[1 + \exp(\lambda c_0)(c_0/c - 1)]^{-1}$. Thus when $c = c_0$, $c' = c_0$ so that this group does in fact leave the integral curve $y = c_0e^x$ fixed.

Chapter 3

3.1 A suitable group is $x' = x$, $y' = y + \lambda$, which extends to $\dot{y}' = \dot{y}$ and $\ddot{y}' = \ddot{y}$. An invariant is $u = x$; a first differential invariant is $v = \dot{y}$. Thus $dv/du = dv/dx = \ddot{y}$. Therefore, $1 + v^2 = k^2(a - x)^2(dv/dx)^2$. The variables separate in this equation, and it can be integrated to give $\operatorname{arcsinh} v = -(1/k)\ln(a - x) + \ln c$, where c is a constant of integration. This last equation can be rewritten as $\dot{y} = v = (1/2)[c(a - x)^{-1/k} - (1/c)(a - x)^{1/k}]$, which can be integrated to give $y = \frac{ck}{2(1-k)}(a-x)^{(k-1)/k} + \frac{k}{2c(1+k)}(a-x)^{(k+1)/k} + b$, where b is a second constant of integration.

3.2 A suitable group is $x' = x + \ln\lambda$, $y' = \lambda^{1/2}y$, which extends to $\dot{y}' = \lambda^{1/2}\dot{y}$ and $\ddot{y}' = \lambda^{1/2}\ddot{y}$. An invariant is $u = y^2e^{-x}$; a first differential invariant is $v = \dot{y}/y$. Then

$$\frac{dv}{dx} = \frac{\ddot{y}}{y} - \left(\frac{\dot{y}}{y^2}\right) = \frac{(1/u - v^2)}{3} - v^2 \qquad (1a)$$

and

$$\frac{du}{dx} = 2y\dot{y}e^{-x} - y^2e^{-x} = 2uv - u \qquad (1b)$$

so that

$$\frac{dv}{du} = \frac{1 - 4uv^2}{3u^2(2v - 1)} \qquad (1c)$$

Equating the numerator and denominator of Eq. $(1c)$ simultaneously to zero, we find that $u = 1$ and $v = 1/2$ are the coordinates of the singular point, which corresponds to the solution $y = e^{x/2}$ in view of the definitions of u and v.

3.3 If the differential equation is written $w(x, y, \dot{y}, \ddot{y}) = 0$, the function w must satisfy Eq. $(3.1.2a)$ for both groups. For the stretching group, $\xi = x$, $\eta = y$, $\eta_1 = 0$, $\eta_2 = -\ddot{y}$; for the translation group, $\xi = 1$, $\eta = 1$, $\eta_1 = 0$, $\eta_2 = 0$. Hence,

$$x w_x + y w_y - \ddot{y} w_{\ddot{y}} = 0 \tag{1a}$$

and

$$w_x + w_y = 0 \tag{1b}$$

The characteristic equations of Eq. (1a) are $dx/x = dy/y = d\dot{y}/0 = -d\ddot{y}/\ddot{y}$; three independent integrals are y/x, $\dot{y}$, and $x\ddot{y}$. Thus the most general form of w satisfying Eq. (1a) is $w = F(y/x, \dot{y}, x\ddot{y})$. Then

$$w_x = \frac{-x_1 F_1 + x_3 F_3}{x} \quad \text{and} \quad w_y = \frac{F_1}{x} \tag{2}$$

where $x_1 \equiv y/x$, $x_2 \equiv \dot{y}$, $x_3 \equiv x\ddot{y}$ and $F_1 \equiv \partial F/\partial x_1$, etc. Thus according to Eq. (1b), F obeys the partial differential equation

$$(1 - x_1) F_1 + x_3 F_3 = 0 \tag{3}$$

The characteristic equations of Eq. (3) are $dx_1/(1 - x_1) = dx_2/0 = dx_3/x_3$. Two independent integrals are $x_3(x_1 - 1) = \ddot{y}(y - x)$ and $x_2 = \dot{y}$. Thus

$$w = F[\ddot{y}(y - x), \dot{y}] \tag{4}$$

The differential equation we are seeking can then be written

$$\ddot{y}(y - x) = G(\dot{y}) \tag{5}$$

where G is any function whatever.

3.4 It follows from the given transformation equations that $\dot{y}' = \lambda \dot{y}$ so that $u = x$ is an invariant and $v = \dot{y}/y$ is a first differential invariant. Then we find

$$\frac{dv}{dx} = \frac{\ddot{y}}{y} - \left(\frac{\dot{y}}{y}\right)^2 = -\frac{v}{x} - 1 - v^2 \tag{1}$$

by differentiating and substituting for $\ddot{y}$ from Bessel's equation. For large x Eq. (1) can be written *correct to terms of order* $1/x$ as

$$-\frac{d[v + 1/2x]}{dx} = 1 + \left(v + \frac{1}{2x}\right)^2 \tag{2}$$

which can be integrated to give

$$\arctan\left(v + \frac{1}{2x}\right) = -x + c \tag{3a}$$

or

$$\frac{\dot{y}}{y} = v = -\tan(x - c) - \frac{1}{2x} \tag{3b}$$

where c is a constant of integration. Equation (3b) can be integrated again to give finally

$$y = ax^{-1/2}\cos(x - c) \tag{4}$$

where a is a second constant of integration. Equation (4) describes the asymptotic behavior of the Bessel functions of order zero.

3.5   If we write the differential equation in the form $u(x, y, \dot{y}, \ddot{y}) = 0$, where $u(x, y, \dot{y}, \ddot{y}) = \ddot{y} + P(x)\dot{y} + Q(x)y - R(x)$, then u must satisfy the condition $\xi u_x + \eta u_y + \eta_1 u_{\dot{y}} + \eta_2 u_{\ddot{y}} = 0$. The infinitesimal coefficients of the group are $\xi = 0$, $\eta = Y$, $\eta_1 = \dot{Y}$ and $\eta_2 = \ddot{Y}$. Direct substitution now shows that $\xi u_x + \eta u_y + \eta_1 u_{\dot{y}} + \eta_2 u_{\ddot{y}} = \ddot{Y} + P(x)\dot{Y} + Q(x)Y = 0$ as required.

To determine a differential invariant and a first differential invariant we must find two integrals of the characteristic equations $dx/0 = dy/Y = d\dot{y}/\dot{Y}$. It follows from the first equality that $x = c$, a constant. Substituting this value of x in the second and third terms tells us that we can treat Y and $\dot{Y}$ as constants when integrating the second equality. Then we find that $\dot{y}Y - y\dot{Y} = a$, a second constant. Thus $u = x$ is an invariant and $v = \dot{y}Y - y\dot{Y}$ is a first differential invariant. Differentiating, we find $dv/dx = \ddot{y}Y - y\ddot{Y} = RY - Pv$, which is the desired associated equation of first order.

3.6   The stretching group with its first two extensions is $x' = \lambda x$, $y' = \lambda^{-2}y$, $\dot{y}' = \lambda^{-3}\dot{y}$, $\ddot{y}' = \lambda^{-4}\ddot{y}$. Power-law solutions, if they exist, must then have the form $y = A/x^2$; substitution of this form into the differential equation shows that $A = 1$. An invariant is $u = x^2 y$; a first differential invariant is $v = x^3\dot{y}$.

Then

$$x\frac{dv}{dx} = 3x^3\dot{y} + x^4\ddot{y} = 4v(1-u) \tag{1a}$$

$$x\frac{du}{dx} = 2x^2 y + x^3\dot{y} = 2u + v \tag{1b}$$

so that

$$\frac{dv}{du} = \frac{4v(1-u)}{2u+v} \tag{1c}$$

The loci of zero slope ($dv/du = 0$) are the u-axis ($v = 0$) and the vertical line $u = 1$. The locus of infinite slope ($dv/du = \infty$) is the oblique line $2u + v = 0$. The singular points are the origin O: $(0,0)$ and the point P: $(1, -2)$. The slope of the integral curves is negative in the triangular region bounded by the u-axis, the line $u = 1$, and the line $2u + v = 0$. All integral curves but one emanating from the origin O either intersect the line $u = 1$ or the line $2u + v = 0$; the one exception is the separatrix passing from O to P. Applying l'Hôpital's rule to the right-hand side of Eq. (1c), we find the quadratic equation $m^2 + 2m - 8 = 0$ for the slope m of this separatrix the roots of which are $m = -4$ and $m = 2$. We want the negative root $m = -4$. Then according to theorem 3.4.1, since $m < \beta = -2$, as we approach the singular point P from the origin O along the separatrix, $x \to \infty$. (A similar use of l'Hôpital's rule shows that the slope m of the separatrix at O is zero. Since then $m > \beta$, as we approach the origin O from the singular point P along the separatrix, $x \to 0$.) Therefore, the asymptotic behavior of the family of solutions $y(x)$ corresponding to the separatrix is $y \sim u_P/x^2 = 1/x^2$.

3.7 The reasoning is exactly the same as that used in connection with Eq. (3.3.13). The differential equation for the infinitesimal difference $u = \delta y$ between two neighboring solutions is

$$x\ddot{u} + \dot{u}(4x^2 y - 1) + 4x^2\dot{y}u = 0 \tag{1}$$

We wish to show that if $u(0) > 0$, then $u(x) \geq 0$ for all $x > 0$. If $u(x)$ were negative anywhere, it would have to have a negative minimum somewhere since it starts out positive and approaches zero at infinity. At a negative minimum, $\ddot{u} \geq 0, \dot{u} = 0$ and $u < 0$. These three requirements contradict Eq. (1); thus the assumption that $u(x)$ is negative anywhere is false and $u(x) \geq 0$ for all $x > 0$.

Note that this argument also requires us to know that $\dot{y} < 0$. Since $y(0) > 0$ and $y(\infty) = 0$, $\dot{y}$ must be < 0 somewhere. Now if $\dot{y}$ were ≥ 0 somewhere else, then $\dot{y}$ would have to be 0 somewhere. But when $\dot{y} = 0$, $\ddot{y} = 0$, and the only solution fulfilling these conditions simultaneously is $y = 0$, which contradicts the hypothesis that $y(0) > 0$. Thus $\dot{y} < 0$ everywhere.

3.8 Differentiating u and v with respect to x we find

$$x\frac{dv}{dx} = -\frac{2y^2\dot{y}}{x^2} + \frac{1}{x}\frac{d(y^2\dot{y})}{dx} = -2v - \dot{y} = -2v - \frac{v}{u^2} \qquad (1a)$$

$$x\frac{du}{dx} = -\frac{y}{x} + \dot{y} = -u + \frac{v}{u^2} \qquad (1b)$$

so that

$$\frac{dv}{du} = \frac{v(1 + 2u^2)}{u^3 - v} \qquad (1c)$$

Positive, decreasing solutions must have $u > 0$ and $v < 0$ and so correspond to curves in the fourth quadrant of the (u, v)-plane. In the neighborhood of the origin, where u and v are small, the term $2u^2$ may be neglected compared to 1. Then Eq. (1c) becomes

$$\frac{dv}{du} = \frac{v}{u^3 - v} \qquad (2)$$

This equation is not immediately integrable, but we may study with profit the three mutually exclusive possibilities: (i) $u^3 \ll v$; (ii) $u^3 \sim v$; and (iii) $u^3 \gg v$ as u and v approach the origin along an integral curve. The first assumption leads to $v = -u$, which is self-consistent, i.e. does not contradict itself when u and v are small. Assumption (ii), which means $v = ku^3$, where k is a constant, leads to $v = ku/(1 - k)$, and so does contradict itself.

The third assumption leads to $v = c \exp[-1/(2u^2)]$, where c is a constant and is also self-consistent.

On the integral curve that enters the origin with a slope of -1, $y^2 \dot{y}/x^2 = -y/x$ when u and v are small. Thus when u and v are small, $y^2 + x^2 = a^2$, a constant. Furthermore, when u is small, y must be small (since x cannot exceed a). Thus the solutions $y(x)$ that correspond to the integral curve entering the origin with a slope of -1 all vanish at some finite value of $x = a$ and are zero thereafter. In the neighborhood of $x = a$, $y = (a^2 - x^2)^{1/2}$, and this form can be used to compute starting values for a numerical integration backwards from $x = a$ towards $x = 0$.

3.9 The stretching group with its first two extensions is $x' = \lambda x$, $y' = \lambda^{-2} y$, $\dot{y}' = \lambda^{-3} \dot{y}$, $\ddot{y}' = \lambda^{-4} \ddot{y}$. Power-law solutions, if they exist, must then have the form $y = A/x^2$; substitution of this form into the differential equation shows that $A = 6$. An invariant is $u = x^2 y$; a first differential invariant is $v = x^3 \dot{y}$. Then

$$x \frac{dv}{dx} = 3x^3 \dot{y} + x^4 \ddot{y} = 3v + \frac{u^2}{3} - \frac{uv}{3} \qquad (1a)$$

$$x \frac{du}{dx} = 2x^2 y + x^3 \dot{y} = 2u + v \qquad (1b)$$

so that

$$\frac{dv}{du} = \frac{9v + u^2 - uv}{3(2u + v)} \qquad (1c)$$

Positive, decreasing solutions correspond to $u > 0$, $v < 0$, i.e. to the fourth quadrant of the direction field of Eq. (1c). The locus of zero slope ($dv/du = 0$) in the fourth quadrant is the curve $v = u^2/(u - 9)$. The locus of infinite slope ($dv/du = \infty$) is the oblique line $2u + v = 0$. The singular points are the origin O: $(0, 0)$ and the point P: $(6, -12)$. The slope of the integral curves is negative in the lenticular region bounded by these two loci. All integral curves but one emanating from the origin O either intersect one locus or the other; the one exception is the separatrix passing from O to P. Applying l'Hôpital's rule to the right-hand side of Eq. (1c), we find the quadratic equation $m^2 + m - 8 = 0$ for the slope m of this separatrix. We want the negative root $m = -[1 + \sqrt{(33)}]/2$. Then according to theorem 3.4.1, since $m < \beta = -2$, as we approach the singular point P from the origin O along the separatrix, $x \to \infty$. (Since at O, $m > \beta$, as we approach the origin O from the singular point P along the separatrix, $x \to 0$.) Therefore, the asymptotic behavior of the family of solutions $y(x)$ corresponding to the separatrix is $y \sim u_P/x^2 = 6/x^2$.

To prove ordering, we consider the differential equation for the infinitesimal difference $u = \delta y$ between two neighboring solutions, namely

$$3\ddot{u} + xy\dot{u} + (x\dot{y} - 2y)u = 0 \qquad (2)$$

We wish to show that if $u(0) > 0$, then $u(x) \geq 0$ for all $x > 0$. If $u(x)$ were negative anywhere, it would have to have a negative minimum somewhere since it starts out positive and approaches zero at infinity. At a negative minimum, $\ddot{u} \geq 0$, $\dot{u} = 0$ and $u < 0$. These three requirements contradict Eq. (2) since the quantity $x\dot{y} - 2y$ is negative for positive, decreasing $y(x)$. Thus the assumption that $u(x)$ is negative anywhere is false and $u(x) \geq 0$ for all $x > 0$.

Knowing the solutions are ordered allows us to employ the results of section 3.3 for stretching groups to find that the asymptotic form of positive, decreasing solutions is given by the power-law solution $y = 6/x^2$.

3.10 A third-order differential equation may be written as $w(x, y, \dot{y}, \ddot{y}, y^{(3)}) = 0$. By introducing the new variables $u = \dot{y}$ and $a = \dot{u} = \ddot{y}$, we may write the third-order differential equation as the set of coupled first-order equations

$$\dot{y} = u \qquad (1a)$$

$$\dot{u} = a \qquad (1b)$$

$$w(x, y, u, a, \dot{a}) = 0 \qquad (1c)$$

The equations (1a–c) determine a *three-parameter* family of curves in the *four-dimensional* space whose coordinates are

x, y, u and a. (Consider, as before, the direction field of the infinitesimal vectors at points (x, y, u, a) having components dx, $dy = \dot{y} \, dx$, $du = \dot{u} \, dx$, $da = \dot{a} \, dx$, where $\dot{a}$ is determined from Eq. (1c). Integrating the three first-order equations (1a–c) involves three constants of integration, which label the integral curves.)

If the original differential equation $w(x, y, \dot{y}, \ddot{y}, y^{(3)}) = 0$ is invariant to a one-parameter group G, the transformations of G, twice extended, carry each of these curves into another. Thus the transformations of G partition the curves into one-parameter subfamilies, all the curves of which map into one another. Thus each subfamily is invariant.

These subfamilies, which form a two-parameter collection, can be denoted by two equations of the form

$$\phi(x, y, u, a, \alpha) = 0 \quad \text{and} \quad \psi(x, y, u, a, \beta) = 0 \quad (2)$$

where α and β are the two parameters that label the subfamilies. Since the subfamilies are invariant to G, the functions ϕ and ψ must satisfy the equations

$$\xi(x, y) \, \phi_x + \eta(x, y) \, \phi_y + \eta_1(x, y, u) \, \phi_u + \eta_2(x, y, u, a) \, \phi_a = 0 \tag{3a}$$

$$\xi(x, y) \, \psi_x + \eta(x, y) \, \psi_y + \eta_1(x, y, u) \, \psi_u + \eta_2(x, y, u, a) \, \psi_a = 0 \tag{3b}$$

The characteristic equations are

$$\frac{dx}{\xi(x, y)} = \frac{dy}{\eta(x, y)} = \frac{du}{\eta_1(x, y, u)} = \frac{da}{\eta_2(x, y, u, a)} \tag{4}$$

If p, q and r are, respectively, an invariant, a first differential invariant and a second differential invariant of G, we see that ϕ and ψ must be functions of p, q and r only. Thus

$$\phi = E(p, q, r, \alpha) = 0 \quad \text{and} \quad \psi = F(p, q, r, \beta) = 0 \quad (5)$$

If we eliminate r between these equations, we find that the invariant subfamilies are represented by a relation of the form

$$H(p, q, \alpha, \beta) = 0 \tag{6}$$

Now the relation (6) represents a two-parameter family of curves in (p, q)-space and thus is equivalent to a second-order differential equation. Therefore, if we adopt the invariant p of G and the first differential invariant q of G as new variables, the third-order differential equation $w(x, y, \dot{y}, \ddot{y}, y^{(3)}) = 0$ reduces to a second-order equation in p and q.

It should be clear that this reasoning is capable of generalization to any order of the original differential equation. Thus the introduction of an invariant and a first differential invariant as new variables always reduces the order of an invariant differential equation by one.

3.11 If $\xi = x^2$ and $\eta = xy$, $\eta_1 = y - x\dot{y}$, and $\eta_2 = -3x\ddot{y}$. Then Eqs. (3.1.2b) are

$$\frac{dx}{x^2} = \frac{dy}{xy} = \frac{d\dot{y}}{y - x\dot{y}} = \frac{d\ddot{y}}{-3x\ddot{y}} \tag{1}$$

The equality of the first and second terms gives, upon integration, $y = cx$, where c is a constant. If we substitute cx for y, the equality of the second and third terms gives upon integration $a = x(c - \dot{y}) = y - x\dot{y}$, where a is a second constant. The equality of the first and last terms gives upon integration $x^3\ddot{y} = b$, another constant. Thus the most general solution of Eq. (3.1.2a) can then be written

$$x^3\ddot{y} = F\left(\frac{y}{x}, y - x\dot{y}\right) \tag{2}$$

To reduce the order of the equation, we introduce the invariant $u = y/x$ and the first differential invariant $v = y - x\dot{y}$. Then

$$x^2\frac{du}{dx} = -y + x\dot{y} = -v \tag{3a}$$

$$x^2\frac{dv}{dx} = -F(u, v) \tag{3b}$$

so that

$$\frac{dv}{du} = \frac{F(u, v)}{v} \tag{3c}$$

3.12 *x absent:* The group is $y' = y$, $x' = x + \lambda$ with the obvious extensions $\dot{y}' = \dot{y}$, $\ddot{y}' = \ddot{y}$. An invariant is $u = y$ and a first differential invariant is $v = \dot{y}$. Then

$$x\frac{dv}{dx} = \ddot{y} = f(y, \dot{y}) = f(u, v) \tag{1a}$$

$$x\frac{du}{dx} = \dot{y} = v \tag{1b}$$

so that

$$\frac{dv}{du} = \frac{f(u, v)}{v} \tag{1c}$$

y absent: the group is $x' = x$, $y' = y + \lambda$ with the obvious extensions $\dot{y}' = \dot{y}$, $\ddot{y}' = \ddot{y}$. An invariant is $u = x$ and a first differential invariant is $v = \dot{y}$. Then

$$\frac{dv}{du} = \ddot{y} = f(u, v) \tag{2}$$

3.13 G_1 is the group $x' = x$, $y' = y + \lambda$ for which $\xi = 0$, $\eta = 1$, $\eta_1 = 0$ and $U_1 = \partial/\partial y$. G_2 is the group $x' = \lambda x$, $y' = y$ for which $\xi = x$, $\eta = 0$, $\eta_1 = -\dot{y}$ and $U_2 = x\,\partial/\partial x - \dot{y}\,\partial/\partial \dot{y}$. Since U_1 involves variables independent of those of U_2, it is clear at once that U_1 and U_2 commute. An invariant and a first differential invariant of G_1 are $p = x, q = \dot{y}$. The associated equation in p and q is

$$\frac{dq}{dp} = \frac{dq}{dx} = \ddot{y} = \frac{1 - x\dot{y}}{x^2} = \frac{1 - pq}{p^2} \tag{1a}$$

or

$$(1 - pq)\,dp - p^2\,dq = 0 \tag{1b}$$

This equation is invariant to G_2. A short calculation (cf. Eq. (3.7.6)) shows that

$$\Xi(p) = U_2 p = \left(x\frac{\partial}{\partial x} - \dot{y}\frac{\partial}{\partial \dot{y}}\right)x = x = p \tag{2a}$$

$$H(p, q) = U_2 q = \left(x\frac{\partial}{\partial x} - \dot{y}\frac{\partial}{\partial \dot{y}}\right)\dot{y} = -\dot{y} = -q \tag{2b}$$

Using the infinitesimal coefficients Ξ and H, we find that Lie's integrating factor for Eq. (1b) is

$$\mu^{-1} = \Xi(p)(1 - pq) + H(p, q)(-p^2) = p \qquad (3)$$

Using this integrating factor, we integrate Eq. (1b) and find

$$-\ln c = \ln p - pq = \ln x - x\dot{y} \qquad (4)$$

where c is a constant of integration. The differential equation (4) is invariant to G_1 which could be used to help solve it; but since its variables separate easily, we solve it directly as follows:

$$dy = \ln(cx)\frac{dx}{x} = \ln(cx)d[\ln(cx)] \qquad (5a)$$

$$y = \frac{[\ln(cx)]^2}{2} + a \qquad (5b)$$

where a is another constant of integration.

3.14 Noether's first integral (3.10.5) is $N = -(x\dot{y})^2/2 + y + \beta(x\dot{y} - \ln x)$. Since the differential equation does not involve β (this is a coincidence), the value of β is arbitrary. This means that the quantities $-(x\dot{y})^2/2 + y$ and $x\dot{y} - \ln x$ must separately be constant along a trajectory in order for N to be constant along the trajectory. Thus

$$x\dot{y} - \ln x = \ln c \qquad \text{and} \qquad -\frac{(x\dot{y})^2}{2} + y = a \qquad (1)$$

where a and c are constants. If we eliminate $\dot{y}$ between these last two equations, we obtain $y = [\ln(cx)]^2/2 + a$, as in the last problem.

3.15 According to Eq. (1.5.5)

$$\eta_1 = \eta_x + (\eta_y - \xi_x)\dot{y} - 2\dot{y}^2\xi_y = \dot{y}\left(\frac{d\eta}{dy} - \frac{d\xi}{dx}\right) \qquad (1)$$

since by hypothesis $\xi = \xi(x)$ and $\eta = \eta(y)$. The characteristic equations (1.7.2b), which in this case take the form,

$$\frac{dx}{\xi} = \frac{dy}{\eta} = \frac{d\dot{y}}{\dot{y}(d\eta/dy - d\xi/dx)} \qquad (2)$$

lead to

$$\frac{d\dot{y}}{\dot{y}} = \frac{dy}{\eta}\left(\frac{d\eta}{dy} - \frac{d\xi}{dx}\right) = \frac{dy}{\eta}\frac{d\eta}{dy} - \frac{dx}{\xi}\frac{d\xi}{dx}$$

$$= \frac{d\eta}{\eta} - \frac{d\xi}{\xi} \tag{3}$$

It follows from Eq. (3) that $\xi\dot{y}/\eta$ is a particular first differential invariant. The most general first differential invariant q is an arbitrary function F of p, a particular invariant, and $\xi\dot{y}/\eta$:

$$q = F\left(p, \frac{\xi\dot{y}}{\eta}\right) \tag{4}$$

Then

$$\left(\frac{\eta}{\xi} - \dot{y}\right)q_{\dot{y}} = \left(\frac{\eta}{\xi} - \dot{y}\right)\frac{\xi}{\eta}F_2 = (1 - q)F_2 \tag{5}$$

which is invariant. Here F_2 signifies the partial derivative of F with respect to its second argument.

3.16 Let us introduce the abbreviations $B \equiv (\eta/\xi - \dot{y})q_{\dot{y}}$ and $w \equiv \eta/\xi - \dot{y}$. The condition that B be invariant is then

$$\xi B_x + \eta B_y + \eta_1 B\dot{y} = 0 \tag{1}$$

or

$$q_{\dot{y}}(\xi w_x + \eta w_y + \eta_1 w_{\dot{y}}) + w(\xi q_{\dot{y}x} + \eta q_{\dot{y}y} + \eta_1 q_{\dot{y}\dot{y}}) = 0 \tag{2}$$

Because q is a first differential invariant

$$\xi q_x + \eta q_y + \eta_1 q_{\dot{y}} = 0 \tag{3}$$

If we differentiate Eq. (3) partially with respect to $\dot{y}$ we find

$$\xi q_{x\dot{y}} + \eta q_{y\dot{y}} + \eta_1 q_{\dot{y}\dot{y}} + q_{\dot{y}}\frac{\partial\eta_1}{\partial\dot{y}} = 0 \tag{4}$$

(Remember ξ and η are functions of x and y only.) Substitution of Eq. (4) into Eq. (2) yields

$$\xi w_x + \eta w_y + \eta_1 w_{\dot{y}} - w\eta_{1\dot{y}} = 0 \tag{5}$$

Now $w_{\dot{y}} = -1$ and $\eta_1 = \eta_x + (\eta_y - \xi_x)\dot{y} - 2\dot{y}^2\xi_y$. Substituting these expressions into Eq. (5) and using the definition of w, we find after reduction

$$\xi_y\left(\frac{\eta^2}{2\xi^2} - \frac{2\eta\dot{y}}{\xi} + \dot{y}^2\right) = 0 \tag{6}$$

Because Eq. (6) is an identity in $\dot{y}$, we must have $\xi_y = 0$, that is, ξ depends only on x, as was to be proved.

3.17

(a) In view of the group property of the transformations (1.1.1*a*, *b*), it is sufficient to require invariance of the area under infinitesimal transformation, i.e. to require

$$0 = \frac{\partial}{\partial\lambda}\iint_{C'}dx'\,dy'|_{\lambda=\lambda_0} = \frac{\partial}{\partial\lambda}\iint_C\left[\frac{\partial(x',y')}{\partial(x,y)}\right]dx\,dy|_{\lambda=\lambda_0}$$

$$= \iint_C\frac{\partial}{\partial\lambda}\left[\frac{\partial(x',y')}{\partial(x,y)}\right]_{\lambda=\lambda_0}dx\,dy \tag{1}$$

Since the closed curve C is arbitrary, Eq. (1) implies that

$$\frac{\partial}{\partial\lambda}\left[\frac{\partial(x',y')}{\partial(x,y)}\right]_{\lambda=\lambda_0} = 0 \tag{2}$$

The Jacobian $\partial(x',y')/\partial(x,y)$ is given by

$$\frac{\partial(x',y')}{\partial(x,y)} = X_xY_y - X_yY_x \tag{3}$$

Thus

$$\frac{\partial}{\partial\lambda}\left[\frac{\partial(x',y')}{\partial(x,y)}\right]_{\lambda=\lambda_0} = \xi_x + \eta_y \tag{4}$$

so that the condition sought for is

$$\xi_x + \eta_y = 0 \tag{5}$$

(b) If the orbits comprise the family $u(x, y) = c$, then according to Eq. (1.3.2)

$$\frac{\eta}{\xi} = -\frac{u_x}{u_y} \tag{6}$$

Thus, most generally,

$$\xi = u_y G(x, y) \qquad \text{and} \qquad \eta = -u_x G(x, y) \tag{7}$$

where $G(x, y)$ is an arbitrary function of x and y. If we substitute ξ and η from Eq. (7) into Eq. (5), we find

$$u_y G_x - u_x G_y = 0 \tag{8}$$

which is the condition for the functional dependence of G on u: $G(x, y) = g(u)$, where $g(u)$ is an arbitrary function of u. Therefore,

$$\xi = u_y g(u) = F_y(u) \qquad \text{and} \qquad \eta = -u_x g(u) = -F_x(u) \tag{9}$$

where $F(u) = \int g(u)\, du$ is an arbitrary function of u.

Chapter 4

4.1 The measured relationship of the temperatures far from the heated end is $c \sim z^a t^b$. Since this is the same in all experiments, the engineer guesses it to be the exceptional solution $c = A z^{L/M} t^{-N/M}$. Thus $L/M = a$ and $N/M = -b$. Then the linear constraint (4.1.2) can be written $\alpha - b\beta = a$. Now $c(0, t) = t^{\alpha/\beta} y(0)$ and $c_z(0, t) = t^{(\alpha-1)/\beta} \dot{y}(0)$. Since α and β differ from experiment to experiment, we must combine in a product the power k of $c(0, t)$ with $c_z(0, t)$ so as to eliminate α and β. Now

$$[c(0, t)]^k c_z(0, t) = t^{[(k+1)\alpha-1]/\beta} [y(0)]^k \dot{y}(0) \tag{1}$$

Now since the linear constraint $\alpha - b\beta = a$ can be written $(\alpha/a - 1)/\beta = b/a$, we see that $k + 1$ must be $1/a$. Thus

$$[c(0, t)]^{(1-a)/a} c_z(0, t) = t^{b/a} [y(0)]^{(1-a)/a} \dot{y}(0) \tag{2}$$

4.2   The coefficients are $M = 1$, $N = -1$, $L = -2$. The exceptional solution thus must have the form $c = Az^{-2}t$; direct substitution shows that the only nonzero value for A is $A = 6$. If $c = t^{\alpha/\beta} y(z/t^{1/\beta})$, then

$$c_t = t^{\alpha/\beta - 1} \frac{(\alpha y - x\dot{y})}{\beta} \qquad \text{where} \qquad x = \frac{z}{t^{1\beta}} \tag{1a}$$

$$c_{zz} = t^{(\alpha - 2)/\beta} \ddot{y} \tag{1b}$$

Then substituting these values in the partial differential equation we find

$$\beta \ddot{y} + xy\dot{y} - \alpha y^2 = 0 \tag{2}$$

This is the principal differential equation. The associated group is $x' = \lambda x$, $y' = \lambda^{-2} y$, with the extensions $\dot{y}' = \lambda^{-3}\dot{y}$, $\ddot{y}' = \lambda^{-4}\ddot{y}$. Clearly each term in Eq. (2) varies with the same factor of λ^{-4} on transformation; this shows that Eq. (2) is invariant to the associated group.

The invariant $u = x^2 y$ and the first differential invariant $v = x^3 \dot{y}$. Then

$$x\frac{dv}{dx} = 3x^3\dot{y} + x^4\ddot{y} = 3v + \frac{\alpha}{\beta}u^2 - \frac{uv}{\beta} \tag{3a}$$

$$x\frac{du}{dx} = 2u + v \tag{3b}$$

$$\frac{dv}{du} = \frac{3v + (\alpha/\beta)u^2 - uv/\beta}{2u + v} \tag{3c}$$

The singularities of Eq. (3c) are determined by setting the numerator and denominator of the right-hand side separately equal to zero. We find then that these singularities are the origin O: $(0, 0)$ and the point P: $(6, -12)$. The value of the coefficient A in the exceptional solution of the partial differential equation is the same as the coefficient u_P in the exceptional solution of the principal differential equation, namely, 6 in this problem.

When $\alpha = -1/2$ (and $\beta = 3/2$), the last two terms in Eq. (2) become the perfect differential $d(xy^2/2)/dx$. Then we can integrate Eq. (2) to obtain $3\dot{y} + xy^2 = 0$ for the solutions that vanish at infinity. This equation is easily integrated again to give $y = 6/(x^2 + c^2)$, where c is a constant of integration. The family thus has the expected asymptotic behavior $y \sim 6/x^2$.

The family $y = 6/(x^2 + c^2)$ has the differential equation $3\dot{y} + xy^2 = 0$ which when multiplied by x^3 becomes the equation $3v + u^2 = 0$. This result satisfies Eq. (3c) when $\alpha = -1/2$ and $\beta = 3/2$. The curve is a separatrix passing through O and P that separates those integral curves in the fourth quadrant that pass through the locus of infinite slope from those that do not.

4.3 Transforming Eq. (4.2.6) with the transformations (4.1.1a–c), we find

$$\lambda^{-\alpha} c' = (\lambda^{-\beta} t')^{\alpha^\circ/\beta^\circ} y \left[\frac{\lambda^{-1} z'}{(\lambda^{-\beta} t')^{1/\beta^\circ}} \right] \tag{1a}$$

which is the same as

$$c' = \lambda^{(\alpha\beta^\circ - \alpha^\circ\beta)/\beta^\circ} t'^{\alpha^\circ/\beta^\circ} y \left[\left(\frac{z'}{t'^{1/\beta^\circ}} \right) \lambda^{-(\beta^\circ - \beta)/\beta^\circ} \right] \tag{1b}$$

Now since $M\alpha + N\beta = L$ and $M\alpha^\circ + N\beta^\circ = L$, it follows that $(\alpha\beta^\circ - \alpha^\circ\beta)/(\beta^\circ - \beta) = L/M$. If we set $\mu = \lambda^{(\beta^\circ - \beta)/\beta^\circ}$, Eq. (1b) becomes

$$c' = \mu^{L/M} t'^{\alpha^\circ/\beta^\circ} y \left(\frac{x'}{\mu} \right) \tag{2}$$

Now since Eq. (2) also represents a similarity solution of the partial differential equation, the function $\mu^{L/M} y(x'/\mu) \equiv y'(x')$ is also a solution of the principal differential equation. It is easy to see that the function $y'(x')$ is obtained from the function $y(x)$ by the transformations of the associated group $x' = \mu x$, $y' = \mu^{L/M} y$. Thus the image of any solution under the transformations of the associated group is another solution, and therefore the principal differential equation is invariant to the associated group.

4.4 By partially differentiating the similarity solution (4.2.6) with respect to t and z, we find

$$c_t = t^{\alpha/\beta - 1} \frac{(\alpha y - x\dot{y})}{\beta} \qquad \left(x = \frac{z}{t^{1/\beta}} \right) \tag{1a}$$

$$c_z = t^{(\alpha-1)/\beta} \dot{y} \tag{1b}$$

$$(cc_z)_z = t^{(2\alpha-2)/\beta} \frac{d(y\dot{y})}{dx} \tag{1c}$$

so that the principal ordinary differential equation is

$$\beta \frac{d(y\dot{y})}{dx} + x\dot{y} - \alpha y = 0 \tag{2}$$

The linear constraint (4.1.2) is $\alpha + \beta = 2$. The invariant $u = x^{-2}y$ and the first differential invariant $v = x^{-1}\dot{y}$. Then

$$x\frac{dv}{dx} = -v + \ddot{y} = -v + \frac{\alpha}{\beta} - \frac{v}{\beta u} - \frac{v^2}{u} \tag{3a}$$

$$x\frac{dv}{dx} = -2u + v \tag{3b}$$

$$\frac{dv}{du} = \frac{-v + \alpha/\beta - v/(\beta u) - v^2/u}{-2u + v} \tag{3c}$$

When $\alpha = -1$ and $\beta = 3$, the principal equation (2) becomes

$$3\frac{d(y\dot{y})}{dx} + x\dot{y} + y = 0 \tag{4}$$

The last two terms in Eq. (4) are the perfect differential $d(xy)/dx$ so that Eq. (4) can be integrated at once. Integrating we find $3y\dot{y} + xy = 0$ for those solutions that vanish at infinity. Thus $\dot{y} = -x/3$, which can be integrated again to yield $y = (a^2 - x^2)/6$, where a^2 is a constant of integration. The constant a can be found from the conservation condition $\int_{-\infty}^{+\infty} c(z,t)\,dz = 1$ which takes the form $\int_{-\infty}^{+\infty} y(x)\,dx = 1$ when $\alpha = -1$; it is $a = (9/2)^{1/3}$.

The family of solutions $y = (a^2 - x^2)/6$ corresponds to the curve $v = -1/3$ in the (u, v)-plane, as can easily be seen by multiplying the differential equation of the family $\dot{y} = -x/3$ by x^{-1}. The point Q in the (u, v)-plane corresponding to $x = 0$ is thus the point $(\infty, -1/3)$ (remember, $u = x^{-2}y$ and $y(0) > 0$). The point P in the (u, v)-plane corresponding to $y = 0$ is thus the point $(0, -1/3)$. When $y = 0$, $x = a$ and $\dot{y} = -a/3$ (remember, $v = x^{-1}\dot{y}$ and $v_P = -1/3$).

When $\alpha = 1/2$ and $\beta = 3/2$, the reduced equation (3c) has three singularities, the origin O: $(0, 0)$, the point R: $(-1/6, -1/3)$ and the point P: $(0, -2/3)$. The singularity P can thus correspond to the conditions $y = 0$, $x = a$ and $\dot{y} = -2a/3$, which can serve as the starting point of a backward numerical integration of the principal differential equation towards the origin. For x larger than a, y could be taken as zero, thereby achieving a positive, decreasing solution that vanished at or beyond $x = a$. Only one numerical integration would have to be undertaken, the remaining members of the family of solutions being obtained from the calculated one by scaling with the associated group (4.3.1)

4.5 By partially differentiating the similarity solution (4.2.6) with respect to t and z, we find

$$c_t = t^{\alpha/\beta-1} \frac{(\alpha y - x\dot{y})}{\beta} \qquad \left(x = \frac{z}{t^{1/\beta}}\right) \tag{1a}$$

$$c_z = t^{(\alpha-1)/\beta} \dot{y} \tag{1b}$$

$$(c^n c_z)_z = t^{[(n+1)\alpha-2]/\beta} \frac{d(y^n \dot{y})}{dx} \tag{1c}$$

so that the principal ordinary differential equation is

$$\beta \frac{d(y^n \dot{y})}{dx} + x\dot{y} - \alpha y = 0 \qquad n\alpha + \beta = 2 \tag{2}$$

The conservation condition $\int_{-\infty}^{+\infty} c(z, t) \, dz = 1$ requires $\alpha = -1$ so that $\beta = n + 2$. The boundary condition $c(\pm\infty, t) = 0$, $t > 0$ then becomes $y(\pm\infty) = 0$ while the conservation condition becomes $\int_{-\infty}^{+\infty} y(x) \, dx = 1$. When $\alpha = -1$, the last two terms in Eq. (2) are the perfect differential $d(xy)/dx$ so that Eq. (2) can be integrated at once. For solutions that vanish at infinity, we find

$$\beta y^n \dot{y} + xy = 0 \tag{3}$$

which can be integrated again to give

$$y = \left[\frac{n(a^2 - x^2)}{2n + 4}\right]^{1/n} \qquad |x| < a \tag{4a}$$

$$y = 0 \qquad\qquad |x| > a \tag{4b}$$

Here a^2 is a constant of integration. From Eq. (4), we find

$$c(z, t) = t^{-1/(n+2)} \left[\frac{n(a^2 - z^2/t^{2/\beta})}{2n + 4} \right]^{1/n} \qquad |z| < at^{1/\beta}$$

(5a)

$$c(z, t) = 0 \qquad\qquad\qquad\qquad\qquad |z| > at^{1/\beta}$$

(5b)

Note that this solution also satisfies the intial condition $c(z, 0) = 0$, $|z| > 0$.

The constant a may be found by substituting Eq. (4) in the conservation condition $\int_{-\infty}^{+\infty} y(x)\, dx = 1$. We note its value for completeness:

$$a = \left(\frac{2n + 4}{n} \right)^{1/(n+2)} \left\{ \pi^{-1/2} \frac{\Gamma[(3n + 2)/(2n)]}{\Gamma[(n + 1)/n]} \right\}^{n/(n+2)}$$

(6)

where Γ is the gamma function.

4.6 If $M \neq 0$, the associated group (4.3.1) can also be written $y' = \mu y$, $x' = \mu^{M/L} x$ if we set $\mu = \lambda^{L/M}$. Now if we let $M \to 0$, we find that the associated group has the form $y' = \mu y$, $x' = x$.

The principal differential equation of the linear diffusion equation is $2\ddot{y} + x\dot{y} - \alpha y = 0$, where $x = z/t^{1/2}$ and $y = c/t^{\alpha/2}$. It is clearly invariant to the associated group just found.

The infinitesimal transformations X and X_* of section 4.3 are given by Eqs. (4.3.2a, b) with $\beta = \beta_* = 2$. The quantities x and y are invariants of X; direct computation shows that $X_* x = 0$ and $X_* y = (\alpha_* - \alpha)y$. The transformations of the group G_* generated in (x, y)-space by X_* are obtained by solving the differential equations $dx/0 = dy/y = d\mu/\mu$ and are $y' = \mu y$, $x' = x$, as was to be shown.

4.7 $\alpha = 0$: the principal differential equation can be written $\ddot{y}/\dot{y} = -x/2$, which can be integrated twice to give $\dot{y} = -c \exp(-x^2/4)$ and $y = c \int_x^\infty \exp(-x^2/4)\, dx = c\pi^{1/2} \operatorname{erfc}(x/2)$, where erfc is the complementary error function and c is a constant of integration.

$\alpha = -1$: the principal differential equation can be written $2\ddot{y} + d(xy)/dx = 0$, which can be integrated twice to give $\dot{y}/y = -x/2, y = c\exp(-x^2/4)$, where c is a constant of integration.

$\alpha = 1$: the *linear* principal differential equation is now $2\ddot{y} + x\dot{y} - y = 0$, which has the particular solution $y = x$. It is noted in problem 3.5 that the differential equation is invariant to the group G whose coefficients are $\xi = 0$ and $\eta = $ the particular solution $= x$. The invariants of this group are $u = x$ and $v = x\dot{y} - y$ and reduce the principal differential equation to $dv/dx = -xv/2$. This last equation can be integrated to give $x\dot{y} - y = v = -c\exp(-x^2/4)$, where c is a constant of integration. This last differential equation is also invariant to G; Eqs. (2.5.6a, b) then tell us that use of the new dependent variable $s = y/x$ separates the variables s and x: x^2 $(ds/dx) = -c\exp(-x^2/4)$. Then one more integration shows that $y = c[\exp(-x^2/4) - (\pi^{1/2}x/2)\,\mathrm{erfc}(x/2)]$.

4.8 Yes! The new condition means that $u_z(0,t) < 0$ for $0 < t < T$. The argument of section 4.5 goes through unchanged until the very end where it must be shown that v cannot attain its smallest value somewhere on the side OA. Now if it did, because $v_z(0,t) < 0$ there would be yet smaller values inside OABC, a contradiction.

4.9 The invariants of the family of groups are $x = ze^{-t/\beta}$ and $y = ce^{-2t/\beta}$. The similarity solutions then take the form $c = e^{2t/\beta}y(ze^{-t/\beta})$. Substituting this form into the partial differential equation, we find the principal ordinary differential equation $\beta d(y\dot{y})/dx + x\dot{y} - 2y = 0$, where $x = ze^{t\beta}$.

The infinitesimal transformations X and X_* of section 4.3 now have the form

$$X = 2c\frac{\partial}{\partial c} + \beta\frac{\partial}{\partial t} + z\frac{\partial}{\partial z} \tag{1a}$$

$$X_* = 2c\frac{\partial}{\partial c} + \beta_*\frac{\partial}{\partial t} + z\frac{\partial}{\partial z} \tag{1b}$$

The quantities x and y are invariants of X; direct computation shows that

$$X_* x = \left(1 - \frac{\beta_*}{\beta}\right) x \quad \text{and} \quad X_* y = 2\left(1 - \frac{\beta_*}{\beta}\right) y \quad (2)$$

The transformations of the group G_* generated in (x, y)-space by X_* are obtained by solving the differential equations $dx/x = dy/2y = d\mu/\mu$ and are $y' = \mu^2 y$, $x' = \mu x$. The principal differential equation found above is invariant to this associated group.

The boundary condition $c(0, t) = e^t$ requires $\beta = 2$.

4.10 The group is the mixed stretching–translation group $c' = c + \alpha \ln \lambda$, $t' = \lambda^\beta t$, $z' = \lambda z$, where $\alpha + \beta = 2$. The invariants of the family are

$$x = z/t^{1/\beta} \quad \text{and} \quad y = c - \left(\frac{\alpha}{\beta}\right) \ln t \quad (1)$$

so that the similarity solutions have the form

$$c = \left(\frac{\alpha}{\beta}\right) \ln t + y\left(\frac{z}{t^{1/\beta}}\right) \quad (2)$$

where the function $y(x)$ is yet to be determined. The infinitesimal transformations X and X_* of section 4.3 now have the form

$$X = \alpha \frac{\partial}{\partial c} + \beta t \frac{\partial}{\partial t} + z \frac{\partial}{\partial z} \quad (3a)$$

$$X_* = \alpha_* \frac{\partial}{\partial c} + \beta_* t \frac{\partial}{\partial t} + z \frac{\partial}{\partial z} \quad (3b)$$

The quantities x and y are invariants of X; direct computation shows that

$$X_* x = \left(1 - \frac{\beta_*}{\beta}\right) x \quad (4a)$$

$$X_* y = \left(\alpha_* - \frac{\alpha \beta_*}{\beta}\right) = 2\left(1 - \frac{\beta_*}{\beta}\right) \quad (4b)$$

The transformations of the associated group G_* generated in (x, y)-space by X_* are obtained by solving the differential equations $dx/x = dy/2 = d\mu/\mu$ and are $y' = y + 2\ln\mu$, $x' = \mu x$. This group has the first extension $\dot{y}' = \mu^{-1}\dot{y}$.

The boundary condition $c(0, t) = a, t > 0$, requires that $\alpha = 0$. The similarity solution then has the form $c = y(z/t^{1/\beta})$.

The quantity $y(\infty) = b$ depends only on the quantities $y(0) = a$ and $\dot{y}(0)$, which for convenience we abbreviate with the letter s. Thus $b = F(a, s)$. Since this relation holds for any solution of the principal differential equation, we also must have $b' = F(a', s')$. Now because the points $x = 0$ and $x = \infty$ transform respectively into the points $x' = 0$ and $x' = \infty$, $b' = b + 2\ln\mu$, $a' = a + 2\ln\mu$ and $s' = \mu^{-1}s$. If we differentiate the equation $b' = F(a', s')$ with respect to μ and set $\mu = 1$, we obtain the linear partial differential equation

$$2 = 2F_a - sF_s \tag{5a}$$

the characteristic equations of which are

$$\frac{dF}{2} = \frac{da}{2} = -\frac{ds}{s} \tag{5b}$$

Two integrals of Eqs. (5b) are $a - F$ and $s^2 e^a$. Since $F = b$, the most general solution of Eq. (5a) is $a - b = f(s^2 e^a)$ where f is an arbitrary function.

4.11 According to Eqs. (4.7.4),

$$c'_{z'}(0, t') = \lambda^{\delta-1}c_z(0, t) \tag{1a}$$

$$t' = \lambda^{3/2-\delta}t \tag{1b}$$

$$p' = \lambda^{3\delta-3}p \tag{1c}$$

If we set $\mu = \lambda^{3\delta-3}$, these equations become

$$c'_{z'}(0, t') = \mu^{1/3}c_z(0, t) \tag{2a}$$

$$t' = \mu^{(3/2-\delta)/(3\delta-3)}t \tag{2b}$$

$$p' = \mu p \tag{2c}$$

The coefficients M, N and L of the linear constraint (4.1.2) must obey the relations $M/3 + N(3/2-\delta)/(3\delta-3) = L$ for all values of

δ; therefore we must have $N = 0$ and $L/M = 1/3$. This means that $c_z(0, t) \sim p^{1/3}$ and is independent of t. This conclusion also follows from Eq. (4.7.13), which implies that $c_z(0, t) = \dot{y}(0)$, and Eq. (4.7.16).

4.12

(a) The group is $v' = \lambda^\alpha v$, $h' = \lambda^\delta h$, $t' = \lambda^\beta t$, $z' = \lambda z$, with the subsidiary conditions $\alpha + \beta = 1$ and $\delta + 2\beta = 2$.

(b) The similarity solutions have the form $v = t^{\alpha/\beta} U(z/t^{1/\beta})$, $h = t^{\delta/\beta} Y(z/t^{1/\beta})$.

(c) The initial conditions given require $\delta = 0$. Therefore $\beta = 1$ and $\alpha = 0$. Thus the similarity solutions become $v = U(z/t)$, $h = Y(z/t)$.

(d) The associated group is $U' = \mu U$, $Y' = \mu^2 Y$, $x' = \mu x$, where $x = z/t$. Now $x_o = z_o/t$ can only depend on h_o; thus $x_o = E(h_o)$. This relation must be invariant to the associated group since it holds for all solutions. Therefore $x'_o = E(h'_o)$, where $h'_o = \mu^2 h_o$ and $x'_o = \mu x_o$. Differentiating with respect to μ and setting $\mu = 1$, we find $E(h_o) = x_o = 2h_o \dot{E}(h_o)$. This last equation is easily solved to give $x_o = E(h_o) = ch_o^{1/2}$.

(e) Partially differentiating, we find $v_t = t^{-1}(-x\dot{U})$ and $v_z = t^{-1}\dot{U}$ and analogous formulas for h. Then the partial differential equations become

$$(U - x)\dot{Y} + Y\dot{U} = 0 \qquad (1a)$$
$$\dot{Y} + (U - x)\dot{U} = 0 \qquad (1b)$$

In order for these equations to have a nonzero solution, the determinant of the coefficients must vanish; thus

$$(U - x)^2 = Y \qquad (2)$$

Now when $x = x_o$, $U = v = 0$ and $Y = h = h_o$ so that $x_o = h_o^{1/2}$. This is the form expected from part (d). It should be noted that the group-theoretic argument of part (d) cannot determine the constant c.

Eliminating Y between Eqs. (1a) and (2) we find $3\dot{U} = 2$, which we can integrate to obtain

$$U = \tfrac{2}{3}(x - h_o^{1/2}) \tag{3}$$

where the constant of integration has been chosen to satisfy the boundary condition $v = 0$ at $x_o = h_o^{1/2}$. We finally find from Eqs. (3) and (2) that

$$Y = \frac{(x + 2h_o^{1/2})^2}{9} \tag{4}$$

The downstream boundary occurs at $x = -2h_o^{1/2}$, where $h = 0$.

These results are the same as those found in Appendix B, where the same problem has been treated by Riemann's method of characteristics.

Chapter 5

5.1 If $c = y(x)$, $x = z - \alpha t$, then $\ddot{y} = -\alpha \dot{y}$. The general solution of this differential equation is $y = A\exp(-\alpha x) + b$, where A and b are constants of integration. Equation (5.6.3) says $v = -c_t/c_z = -(-\alpha\dot{y})/\dot{y} = \alpha$.

5.2 If $u = U(x)$ and $v = Y(x)$, $x = z - \alpha t$, then Eqs. (4.9.1a, b) become $-\alpha\dot{U} = \dot{Y}$ and $-\alpha\dot{Y} = V^2\dot{U}$. Thus $V^2 = \alpha^2$ so that there can only be traveling-wave solutions if V is constant.

5.3 If $c = y(x)$, $x = z - \alpha t$, then $\ddot{y} = \alpha^2\ddot{y}$. Thus $\alpha^2 = 1$ so that $\alpha = \pm 1$. Furthermore, the equation puts no constraint whatever on $y(x)$, which can thus be any function of x. Taken alone neither the function $y(z - t)$ nor the function $y(z + t)$ can satisfy the initial conditions $c(z, 0) = u(z)$, $c_t(z, 0) = 0$. For $y(z) = c(z, 0) = u(z)$ and $-\alpha\dot{y}(z) = c_t(z, 0) = 0$, which are contradictory unless $u(z)$ is a constant. We therefore try a linear combination

$$c(z, t) = A\,y(z - t) + B\,y(z + t) \tag{1}$$

The initial conditions are then

$$Ay(z) + By(z) = u(z) \tag{2a}$$

$$-A\dot{y}(z) + B\dot{y}(z) = 0 \tag{2b}$$

According to Eq. (2b), $A = B$. If we incorporate the constant A into the function y, we find that $y(z) = u(z)/2$ so that

$$c(z, t) = \frac{u(z - t) + u(z + t)}{2} \tag{3}$$

When $c(z, 0) = 0$, $c_t(z, 0) = v(z)$ we again try a solution of the form (1). The initial conditions are now

$$Ay(z) + By(z) = 0 \tag{4a}$$

$$-A\dot{y}(z) + B\dot{y}(z) = v(z) \tag{4b}$$

Thus $A = -B$ and incorporating B into the definition of y, we have $\dot{y}(z) = v(z)/2$ or $y = (1/2)\int_0^z v(z')\,dz'$. Thus

$$c(z, t) = \frac{1}{2}\left[\int_0^{z+t} v(z')\,dz' - \int_0^{z-t} v(z')\,dz'\right]$$

$$= \frac{1}{2}\int_{z-t}^{z+t} v(z')\,dz' \tag{5}$$

5.4 If $c = y(x)$, $x = z - \alpha t$, then $d(y\dot{y})/dx + \alpha\dot{y} + y(1 - y) = 0$. The invariants of the associated group are $u = y$ and $v = \dot{y}$. Then

$$\frac{dv}{dx} = \ddot{y} = -\frac{v^2 + \alpha v + u(1 - u)}{u} \tag{1a}$$

$$\frac{du}{dx} = \dot{y} = v \tag{1b}$$

$$\frac{dv}{du} = -\frac{v^2 + \alpha v + u(1 - u)}{uv} \tag{1c}$$

The straight line joining P: $(1, 0)$ and Q: $(0, -\alpha)$ is $v = \alpha(u - 1)$. Substituting this straight-line solution in Eq. (1c) we find the latter can only be satisfied if $\alpha = 2^{-1/2}$. Then $\dot{y} = 2^{-1/2}(y - 1)$. Integrating this equation once more, we find $y(x) = 1 - \exp(2^{-1/2}x)$ when $0 < u < 1$ ($x < 0$) and $y(x) = 0$ otherwise ($x > 0$). The constant of integration has been chosen so that the wave front is at $z = 0$ when $t = 0$ and the traveling-wave solution obeys the boundary conditions $y(-\infty) = 1$ and $y(\infty) = 0$.

5.5 The infinitesimal transformation of the family F is

$$X = \alpha \frac{\partial}{\partial z} + \frac{\partial}{\partial t} \tag{1}$$

Suppose the infinitesimal transformation of the group G is

$$Y = \zeta \frac{\partial}{\partial z} + \tau \frac{\partial}{\partial t} + \gamma \frac{\partial}{\partial c} \tag{2}$$

If the commutator of X and Y is a linear combination of them, then, as we have seen in section 3.7, the principal differential equation in the invariants $y = c$ and $x = z - \alpha t$ of F is also invariant to G. Now

$$XY - YX = (\alpha \zeta_z + \zeta_t) \frac{\partial}{\partial z} + (\alpha \tau_z + \tau_t) \frac{\partial}{\partial t} + (\alpha \gamma_z + \gamma_t) \frac{\partial}{\partial c} \tag{3}$$

For this commutator to be a linear combination of $aX + bY$, we must have

$$\alpha \zeta_z + \zeta_t = a\alpha + b\zeta \tag{4a}$$

$$\alpha \tau_z + \tau_t = a + b\tau \tag{4b}$$

$$\alpha \gamma_z + \gamma_t = b\gamma \tag{4c}$$

Since Eqs. (4a–c) are identities in α, these three expressions become the six equations

$$\zeta_z = a \qquad \zeta_t = b\zeta \tag{5a}$$

$$\tau_z = 0 \qquad \tau_t = a + b\tau \tag{5b}$$

$$\gamma_z = 0 \qquad \gamma_t = b\gamma \tag{5c}$$

Several possibilities exist that depend on whether a is zero or not and b is zero or not.

5.6 In order to keep $y = 1$ as the breakpoint between the two forms of the differential equation we need to take $y' = y$. If we choose $u' = \lambda u$ then $\dot{u}' = \lambda \dot{u}$. Next we must have $\alpha' = \lambda \alpha$ since α must transform like $\dot{u}$. Finally γ must transform like $u\dot{u}$ so that $\gamma' = \lambda^2 \gamma$. Now α can only be a function of γ and the functional dependence must be the same for all pairs of these parameters. Thus $\alpha' = f(\gamma')$. Differentiating with respect to λ and setting $\lambda = 1$, we find $\alpha = 2\gamma(df/d\gamma)$. Now since $\alpha = f(\gamma)$, we find the simple differential equation $df/f = d\gamma/(2\gamma)$, the solution to which is $\alpha = f = C\gamma^{1/2}$, where C is a constant of integration that cannot be determined by group-theoretic arguments. The argument goes through without change for Eq. (5.2.2) when $Q(y)$ is replaced by $\gamma Q(y)$.

5.7 If $b = B(x)$, $n = N(x)$, where $x = z - \alpha t$, then

$$-\alpha \frac{dB}{dx} = \frac{d}{dx}\left(\frac{dB}{dx} - \frac{2B}{N}\frac{dN}{dx}\right) \tag{1a}$$

$$\alpha \frac{dN}{dx} = B \tag{1b}$$

The infinitesimal transformations of groups (2) and (3) in the problem statement are, respectively, $X_2 = \alpha\,\partial/\partial z + \partial/\partial t$ and $X_3 = n\,\partial/\partial n + b\,\partial/\partial b$. These operators commute since they involve entirely different variables. This means that the Eqs. (1a, b) must be invariant to the transformations $N' = \lambda N$, $B' = \lambda B$, as they clearly are.

We can integrate Eq. (1a) to obtain

$$-\alpha B = \frac{dB}{dx} - \frac{2B}{N}\frac{dN}{dx} \tag{2}$$

Here the constant of integration has been taken to be zero because when $B = 0$, $dN/dx = B/\alpha = 0$ and $dB/dx = 0$†. Dividing (2) by B separates the variables; a second integration yields

$$N = [1 + e^{-\alpha x}]^{-1} \quad \text{and} \quad B = \alpha^2 e^{-\alpha x}[1 + e^{-\alpha x}]^{-2} \quad (3)$$

The constants of integration have been chosen to make $N(-\infty) = 0$ and $N(\infty) = 1$ and $N(0) = 1/2$.

5.8 The fourth quadrant of the direction field resembles the direction field in figure 5.5.1. There are two straight lines radiating from the point $(1, 0)$ which have slopes $-\kappa_\pm$, the roots of Eq. (5.5.2b). Now an integral curve starting at the origin with slope $-\alpha$ must eventually approach the curve $u = -\kappa_+(y - 1)$ asymptotically; here $\kappa_+ = [\alpha + (\alpha^2 - 4)^{1/2}]/2$. Thus $\dot{y} = -\kappa_+(y - 1)$ so that $y \sim \exp(-\kappa_+ x)$ when $y \gg 1$; this occurs when x is large and negative.

Chapter 6

6.1 The partial differential equation $cc_t = c_{zz}$ is invariant to the family of groups (4.1.1a–c) with the subsidiary condition $\alpha - \beta = -2$. The principal differential equation obeyed by the similarity solutions $c = t^{\alpha/\beta} y(z/t^{1/\beta})$ is $\beta \ddot{y} + y(x\dot{y} - \alpha y) = 0$, $x = z/t^{1/\beta}$. When the boundary and initial conditions are $c(0, t) = 1$, $t > 0$; $c(\infty, t) = 0$; $c(z, 0) = 0$, $0 < z < \infty$, $\alpha = 0$, $\beta = 2$, the similarity solution has the form

$$c = y\left(\frac{z}{t^{1/2}}\right) \quad (1a)$$

and the principal differential equation becomes

$$2\ddot{y} + xy\dot{y} = 0 \quad (1b)$$

---

† According to Eq. (1a), dB/dx must be continuous; for if it were not and had a finite jump anywhere, the right-hand side of Eq. (1a) would be infinite whereas the left-hand side would not. Far ahead and far behind the moving colony $B(x)$ is flat and equals zero. Thus when $B = 0$, $dB/dx = 0$.

When $c(0, t) = F(t)$, we try a solution of the form

$$c(z, t) = F(t)y\left[\frac{z}{p(t)}\right] \tag{2}$$

where $y(x)$ is the solution of Eq. (1b) that satisfies the boundary conditions $y(0) = 1$, $y(\infty) = 0$.

If we integrate the partial differential equation with respect to z from zero to infinity, we find

$$\frac{d}{dt} \int_0^\infty \frac{c^2}{2} dz = -c_z(0, t) \tag{3}$$

If we substitute Eq. (2) into Eq. (3), we find

$$\frac{d}{dt}\left[F^2 p \int_0^\infty \frac{y^2}{2} dx\right] = -\frac{F\dot{y}(0)}{p} \tag{4}$$

Now from Eq. (1b) we find that

$$0 = 2\dot{y}\,\Big|_0^\infty + \int_0^\infty xy\dot{y}\,dx = -2\dot{y}(0) - \int_0^\infty \frac{y^2}{2} dx \tag{5}$$

where the second equality is obtained by an integration by parts. Thus

$$\frac{d}{dt}(2F^2 p) = F/p \quad \text{or} \quad 2F^2 p\frac{d}{dt}(F^2 p) = F^3 \tag{6}$$

Integrating we find

$$p = \frac{\left(\int_0^t F^3\,dt\right)^{1/2}}{F^2} \tag{7}$$

which obeys the condition $p \sim t^{1/2}$ for $t \ll 1$ that follows from the similarity solution corresponding to constant F. Finally, then,

$$c_z(0, t) = \frac{F\dot{y}(0)}{p} = \frac{F^3\dot{y}(0)}{\left(\int_0^t F^3\,dt\right)^{1/2}} \tag{8}$$

6.2 We use as a basis for approximation the similarity solution to the principal differential equation when $\alpha = 1$ (and $\beta = 3$). Then

$$c = t^{1/3} y\left(\frac{z}{t^{1/3}}\right) \tag{1a}$$

$$3\ddot{y} + xy\dot{y} - y^2 = 0 \qquad x = \frac{z}{t^{1/3}} \tag{1b}$$

$$\dot{y}(0) = -1 \qquad y(\infty) = 0 \tag{1c}$$

When $c_z(0, t) = -G(t)$, we try a solution of the form

$$c(z, t) = G(t)\, p(t)\, y\left[\frac{z}{p(t)}\right] \tag{2}$$

As before,

$$\frac{d}{dt} \int_0^\infty \frac{c^2}{2} dz = -c_z(0, t) \tag{3}$$

which now becomes

$$\frac{d}{dt}\left[G^2 p^3 \int_0^\infty \frac{y^2}{2} dx \right] = G \tag{4}$$

Integrating Eq. (1b) from zero to infinity and then integrating by parts, we find

$$\int_0^\infty \frac{y^2}{2} dx = 1 \tag{5}$$

so that

$$\frac{d}{dt}(G^2 p^3) = G \qquad \text{or} \qquad p = \left(\frac{1}{G^2} \int_0^t G\, dt\right)^{1/3} \tag{6}$$

and

$$c(0, t) = y(0)\left(G \int_0^t G\, dt\right)^{1/3} \tag{7}$$

6.3 The infinitesimal transformation of the group $C' = \lambda^\alpha C$, $t' = \lambda^2 t$, $q' = \lambda^{\alpha-1} q$, $R' = \lambda R$ corresponding to the parameter α is

$$X = \alpha C \frac{\partial}{\partial C} + (\alpha - 1)q \frac{\partial}{\partial q} + R \frac{\partial}{\partial R} + 2t \frac{\partial}{\partial t} \tag{1}$$

Independent invariants of X are

$$u = \frac{R}{t^{1/2}} \qquad v = \frac{q}{R^{\alpha-1}} \qquad w = \frac{C}{t^{\alpha/2}} \tag{2}$$

The infinitesimal transformation corresponding to the parameter α_* is

$$X_* = \alpha_* C \frac{\partial}{\partial C} + (\alpha_* - 1)q \frac{\partial}{\partial q} + R \frac{\partial}{\partial R} + 2t \frac{\partial}{\partial t} \tag{3}$$

Direct calculation shows that

$$X_* u = 0 \qquad X_* v = (\alpha_* - \alpha)v \qquad X_* w = (\alpha_* - \alpha)w \tag{4}$$

so that in (u, v, w)-space

$$X_* = (\alpha_* - \alpha)\left(v \frac{\partial}{\partial v} + w \frac{\partial}{\partial w} \right) \tag{5}$$

the invariants of which are u and w/v. Thus most generally $w/v = f(u)$, where f is an arbitrary function. Substituting from Eqs. (2) we find this latter equation reads $C/t^{\alpha/2} = (q/R^{\alpha-1})f(R/t^{1/2})$ and can be rewritten as

$$C = qt^{1/2}\left(\frac{t^{1/2}}{R} \right)^{\alpha-1} f\left(\frac{R}{t^{1/2}} \right) = qt^{1/2} g\left(\frac{R}{t^{1/2}} \right) \tag{6}$$

where g, too, is an arbitrary function.

The boundary condition $\int_0^\infty rc(r, t)\, dr = Rqt$ requires that $\alpha = 0$ in group 1 of the problem statement. The similarity solution now takes the form $c = y(x)$, $x = r/t^{1/2}$. Then the principal differential equation becomes

$$\ddot{y} + \dot{y}\left(\frac{1}{x} + \frac{x}{2} \right) = 0 \tag{7}$$

with the boundary conditions $\int_0^\infty yx\,dx = Rq$ and $y(\infty) = 0$. Equation (7) can be integrated to give

$$\dot{y} = k\frac{\exp(-x^2/4)}{x} \tag{8}$$

where k is a constant of integration. If we integrate $\int_0^\infty yx\,dx = Rq$ by parts, we find $\int_0^\infty x^2\dot{y}\,dx = -2Rq$ from which we can determine that $k = -Rq$. Then one more integration shows that

$$c = y = Rq\int_x^\infty \frac{\exp(-x^2/4)}{x}dx = \left(\frac{Rq}{2}\right)E_1\left(\frac{r^2}{4t}\right) \tag{9}$$

where $E_1(x)$ is the tabulated exponential integral $\int_x^\infty e^{-x}/x\,dx$.

References

[Ba-72] G. I. Barenblatt and Ya B Zeldovich *Annual Reviews of Fluid Mechanics* **4** 285–312 (1972).

[Bl-89] G. W. Bluman and S. Kumei *Symmetries and Differential Equations*, Springer-Verlag, New York, 1989.

[Co-11] Abraham Cohen *An Introduction to the Lie Theory of One-Parameter Groups (with application to the solution of differential equations)*, G. E. Stechert & Co., New York, 1911 (reprinted 1931).

[Co-48] R. Courant and K. O. Friedrichs *Supersonic Flow and Shock Waves*, Interscience, New York, 1948.

[Co-53] R. Courant and D. Hilbert *Methods of Mathematical Physics*, Interscience, New York, 1953.

[Da-62] Harold T. Davis *Introduction to Nonlinear Differential and Integral Equations*, Dover Publications, Inc., New York, 1962.

[Dr-83] Lawrence Dresner *Similarity Solutions of Nonlinear Partial Differential Equations*, Research Notes in Mathematics #88, Pitman Advanced Publishing Program, Marshfield, MA 02050, 1983.

[Dr-90] Lawrence Dresner *The Superfluid Diffusion Equation*, Oak Ridge National Laboratory Report ORNL/TM-11480, June, 1990, available from the National Technical Information Service, US Department of Commerce, 5285 Port Royal Road, Springfield, VA 22161.

[Dr-95] Lawrence Dresner *Stability of Superconductors*, Plenum Press, New York, 1995.

[Fo-33] Lester R. Ford *Differential Equations*, McGraw-Hill Book Co., New York, 1933.

[Go-95] K. S. Govinder and P. G. L. Leach, On the Determination of Non-Local Symmetries, *J. Phys. A: Math. Gen.* **28** 5349–59 (1995).

[Ma-69] B. J. Maddock, G. B. James and W. T. Norris, Superconductive Composites: Heat Transfer and Steady-State Stabilization, *Cryogenics* **9**: 261–273 (1969).

[Mu-89] J. D. Murray *Mathematical Biology*, Springer-Verlag, New York, 1989.

[No-18] Emmy Noether *Invariante Variationsprobleme*, Nachrichten d. Kgl. Jes. d. Wiss., Math.-phys. Klasse **2** 235–57 (1918).

[Ol-86] Peter J. Olver *Applications of Lie Groups to Differential Equations*, Springer-Verlag, New York, 1986.

[Pa-59] R. E. Pattle *Quart. J. Mech. Appl. Math.* **XII** (4) 407–9 (1959).

[Pr-67] M. H. Protter and H. F. Weinberger *Maximum Principles in Differential Equations*, Prentice-Hall, Englewood Cliffs, NJ, 1967.

[Sm-61] V. I. Smirnov *Linear Algebra and Group Theory*, McGraw-Hill, New York, 1961.

[Wy-74] B. G. Wybourne *Classical Groups for Physicists*, Wiley–Interscience, New York, 1974.

Symbols and their Definitions

Some of the symbols used in this book are used consistently throughout and these symbols are defined below. Occasionally, some of these defined symbols are used differently in different places. When these different uses occur throughout the book, the places where they occur are noted in the definitions given below. The different uses are always made clear in the text. Symbols used only once for a specific purpose or to show functional dependence or as arbitrary constants are not defined below but are always defined in the text where they are used. As mentioned in the section entitled *Conventions Used in this Book*, use is made of the subscript notation for partial derivatives. Thus the ordinary diffusion equation $\partial c / \partial t = \partial^2 c / \partial z^2$ is written $c_t = c_{zz}$ for the sake of economy in typesetting. Similarly Newton's notation for the ordinary derivative is often used in place of Leibniz's; thus for the function $y(x)$, $\dot{y} = \mathrm{d}y/\mathrm{d}x$ and $\ddot{y} = \mathrm{d}^2 y/\mathrm{d}x^2$. Furthermore, the prime always denotes the image of a variable under transformation; thus x' is the image of x under transformation.

Roman Symbols

A coefficient in the exceptional solution $y = Ax^\beta$

c dependent variable in chapter 4 [see Eq. (4.1.1a)] and chapter 5 [see Eq. (5.1.1a)]

D diffusion constant

m slope of either separatrix at a saddle point singularity; §3.4

g right-hand side of Eq. (3.4.5b); similarity variable, Eq. (4.9.5)

h right-hand side of Eq. (3.4.5a); similarity variable, Eq. (4.9.5)

L Lagrangian, §3.9; coefficient in the linear constraint Eq. (4.1.2)

M coefficient in the first-order differential equation (2.1.2); coefficient in the linear constraint Eq. (4.1.2)

N coefficient in the first-order differential equation (2.1.2); Noether's first integral, §3.9–3.10; coefficient in the linear constraint Eq. (4.1.2)

p group invariant, chapter 3, §4.10; pressure rise, §4.7; scaling trial function, §6.2

q first differential invariant, chapter 3, §4.10; function appearing in the conservation equation (4.5.1); coefficient in the pulsed-source formula Eq. (4.6.2); scaling trial function, §6.3

Q source term in the diffusion equation (4.6.1)

r radius vector

S coefficient in the conservation equation (4.5.1)

t independent (time) variable

u group invariant, Eq. (1.3.1a), Eq. (1.7.1); $u = \dot{y}$, §3.1–3.2, chapter 5; $u = \delta y$, the infinitesimal difference between two neighboring solutions, Eqs. (3.3.10), (3.3.13), (4.5.2); variable in the wave-propagation equations (4.9.1)

U infinitesimal transformation, Eq. (3.7.1); potential, Eq. (3.9.11); shock speed, §4.11

v auxiliary variable introduced in Eq. (4.5.3); variable in the wave-propagation equations (4.9.1); local velocity of propagation, §5.6

w group invariant, Eq. (3.1.1)

W source function defined in Eqs. (5.2.4) and (5.2.5)

x Cartesian coordinate in the plane; similarity variable, Eq. (4.2.5)

X transformation for x, Eq. (1.1.1a); infinitesimal transformation, Eq. (4.3.2), Eq. (5.1.9a); value at the front of the similarity variable z/t^2, §4.7

y Cartesian coordinate in the plane; similarity variable, Eqs. (4.2.5) and (5.1.7)

Y transformation equation for y, Eq. (1.1.1b)

$\underline{x}$ new variable, Eq. (2.5.1a)

$\underline{y}$ new variable, Eq. (2.5.1b)

z independent (space) variable

Greek Symbols

α	exponent in the c-transformation equation (4.1.1a), chapter 4; coefficient in the z-transformation equation (5.1.1c); chapter 5
β	exponent in the y-transformation equation, chapter 3; exponent in the t-transformation equation (4.1.1b), chapter 4
γ	scale factor used in chapter 5
η	y-coefficient function
η_1	$\dot{y}$-coefficient function in the extended group, Eq. (1.5.5)
η_k	$y^{(k)}$-coefficient function in the extended group, Eq. (1.6.5)
θ	polar angle
λ	group parameter
λ_0	value of λ, the group parameter, for the identity transformation
μ	Lie's integrating factor, Eq. (2.1.4); group parameter in Eqs. (4.9.8) and (5.1.8)
ν	integrating factor of Eq. (2.1.2) given in Eq. (2.3.3); parameter in the Poisson–Boltzmann equation, Eq. (3.6.1)
ξ	x-coefficient function
$\phi(x, y)$	function used to represent a family of curves, Eq. (1.4.1), Eq. (3.7.9a)
$\psi(x, y)$	function used to represent a family of curves, Eq. (1.4.5)

Index

Printed in the United States
by Baker & Taylor Publisher Services